Food Composition Data:
A User's Perspective

Report of a conference held in Logan, Utah, USA, 26–29 March 1985

Edited by
WILLIAM M. RAND, CAROL T. WINDHAM,
BONITA W. WYSE, and VERNON R. YOUNG

 The United Nations University

Food and Nutrition Bulletin Supplement 12

Supported in part by the United States National Institutes of Health,
Contract No. N01-CN-45168

The United Nations University's International Food Data Systems Project (INFOODS) began
as a comprehensive effort to improve and make available food composition data from and for
all countries of the world. One of the first steps taken by INFOODS was to convene a workshop
to identify and discuss the various purposes for which food composition data are needed and
the ways in which they are currently being produced, maintained, and distributed. This volume
contains the background papers prepared for that workshop and the results of the subsequent
discussion on what needs to be done.

Data on the nutrient content of foods are critical in many areas – health assessment, the
formulation of appropriate institutional and therapeutic diets, food and nutrition training,
epidemiological research, etc. For many parts of the world, however, such data are lacking
or inaccessible, incompatible, and incomplete. This volume offers a careful assessment of what
is needed, the extent to which these needs are currently being met, and the essential long-term
solutions to the existing problems. It also formulates the goal of the UNU–INFOODS project –
to ensure that eventually anyone around the globe will be able to obtain adequate and reliable
food composition data.

The United Nations University
Toho Seimei Building, 15-1 Shibuya 2-chome, Shibuya-ku, Tokyo 150, Japan
Tel.: (03) 499-2811   Telex: J25442   Cable: UNATUNIV TOKYO

Printed in the United Kingdom at The Bath Press, Avon

WHTR-10/UNUP-633
ISBN 92-808-0633-5
United Nations Sales No. E.87.III.A.2
02000 P

Contents

Foreword

The United Nations University was founded by the General Assembly of the United Nations in 1975 to help solve the pressing global problems of human survival and welfare through the instruments of scholarship – research, training, and dissemination of knowledge. From the start, food and nutrition have been major areas of involvement, with the funding of research into human requirements, the advanced training of workers from developing countries in various fields of nutrition and food science, and the sponsoring of workshops and symposia on a range of related subjects.

It was realized early that good, readily available data on the composition of foods consumed by man were essential for all of these activities, as well as for many others in the international world of trade, research, and assistance. The amount, quality, and availability of these data vary between and within countries and regions. Moreover, for no area is there a body of data that users feel is totally adequate for their needs – in that accurate, accessible data simply do not exist for most of the foods of the world. In early 1983 the United Nations University organized a small gathering of international experts in Bellagio, Italy, to assess the status and problems of food composition data, explore what could be done, and plan its execution. This meeting, attended by representatives of international agencies and scientific unions, government agencies, and individuals from the academic community, was sponsored by the United States government, the food industry, and private foundations. The group recommended the formation of INFOODS (the International Network of Food Data Systems) to stimulate and co-ordinate efforts to improve the status of food composition data around the world. Work got formally under way in the summer of 1984 with the establishment of a secretariat at the Massachusetts Institute of Technology, funded primarily by the United States government (NCI, NHLBI, USDA, and FDA), but with additional funding from industry and private foundations, and administrative assistance from the United Nations University.

One of the first activities that INFOODS felt necessary was a careful assessment of the needs of the food composition data user community, of whether these needs were or were not being satisfied by the present data systems and what the users saw as solutions to their problems. To effect this, INFOODS designed a conference on

Users and Needs of Food Composition Data to bring together a number of individuals involved with different aspects of food composition data; it was planned as a collaboration between the INFOODS secretariat and Utah State University, with the major responsibility for organization being assumed by Professor Gaurth Hansen and his associates, Professors Carol Windham and Bonita Wyse of Utah State University. The conference took place from 26 to 29 March 1985 on the campus of Utah State University in Logan, Utah, and the major goal was to discuss and document the current situation and thinking about food composition data systems. As this document attests, the conference was very successful, both in defining the needs of the user community and in indicating how INFOODS should try to satisfy them. Although INFOODS' ultimate goal of making food composition data complete, accurate, and available is still a long way off, this conference has made an auspicious start.

Nevin S. Scrimshaw,
Director,
Development Studies Division,
United Nations University

Preface

INFOODS is a collaboration of individuals and organizations interested in food composition data. It is co-ordinated by a secretariat based at the Massachusetts Institute of Technology, internationally supported by the United Nations University, and funded primarily by the United States agencies of the National Cancer Institute, the National Heart, Lung and Blood Institute, the Food and Drug Administration, and the US Department of Agriculture. Thanks are given to all these agencies for their assistance in holding this meeting.

Additional thanks are due to Professor Gaurth Hansen and his colleagues, Professors Carol Windham and Bonita Wyse, and their staffs at Utah State University for their efforts in the organization and arrangements, and in making the conference enjoyable as well as productive.

This report details the ideas, discussions, and recommendations of a very diverse group of individuals, drawn from quite different disciplines and situations. Since the meeting, the secretariat of INFOODS has been using the results of these discussions to organize its priorities, and to help start new activities. An indication of the importance of this meeting is that steps have been taken to implement each of their recommendations. Thus, the INFOODS secretariat, and indeed the community, owe special thanks to the participants, who gave of their time and ideas to meet and help chart the course of INFOODS.

William M. Rand,
Executive Secretary,
INFOODS

Executive Summary

Food Composition Data Are Important

Food is a major component of man's environment. Data on what is actually in foods are critical to important activities of a great variety of individuals and groups including those involved with epidemiological research into disease patterns, government regulation formulation and enforcement, health assessment of individuals and populations, and national and international trade in foods.

Food Composition Data Are Currently Inadequate and Getting Worse

Current activities involved with the gathering and compiling of food composition data are for the most part isolated and often produce inaccessible, incompatible, inconsistent, and redundant information. Additionally, there are large gaps in the available data on what is in foods; data on many biologically important components of many commonly consumed foods simply do not exist. Moreover, very little information exists on the variability of food components and the attendant problems are almost totally neglected by data compilers and users.

Most food data bases are compiled for specific usages and often are unsuitable for other purposes. Thus, many of the data that do exist are difficult to find and use because of the limiting ways in which they are gathered, stored, and described. Interest in food components is expanding rapidly, as are the varieties of foods that are available and consumed. The need for access to more data on more foods is far outstripping the amount of data that are being made available.

There Is a Need for Co-ordinated Activity in the Field

The importance of food composition data and the relatively limited resources that are available necessitate a co-ordination of activity in the area. It is recommended that INFOODS pursue this goal by:

1. Serving as a focal point as it strives to set up a network of global communication between workers in the field.
2. Formulating guidelines for the various aspects of gathering, compiling, communicating, and using food composition data. This, of necessity, will include guidelines for methodologies for data gathering, data description, data storage and retrieval, data interchange, and data manipulation.
3. Pursuing the appropriate involvement of modern computer and information systems technology within the field of food composition data. Any such involvement must recognize that flexibility is a key element and that work in this area must be primarily directed towards providing the users with better tools.

Report and Recommendations of the Conference

INFOODS was organized after a meeting of experts concerned with the unsatisfactory state of food composition data at the international level [9]. It was formed in response to a perceived sense that there were major problems in the area of food composition data, and therefore one of its first tasks was to determine the status of food composition data, and to decide which users' needs were not being met and what should and could be done. Thus, INFOODS helped organize a meeting involving a number of individuals directly concerned with food composition data. This was held on the campus of Utah State University in late March 1985, with the major goals being to define and document the current situation, and to make proposals for the future. (See the appendix for a list of the participants.) The format of the meeting included presentation of papers followed by discussion by the full group; small writing groups then prepared the material on which this report was based. After the meeting the authors were given the opportunity to revise their papers, and these papers follow this report of the conference. The report itself is divided into sections that introduce the papers (section 1), discuss the users themselves and their needs (section 2), examine the data that are available (section 3), explore the design of an ideal food composition data system (section 4), and finally make recommendations to INFOODS for future activities (section 5).

1. INTRODUCTION

The papers presented can be divided into five general areas. The first three papers, each by a member of the INFOODS staff, deal with the background and general context of the meeting. Vernon Young (paper 1) discusses the forces that led to the formation of INFOODS, arguing that it is now time for the field of food composition data to be drawn together and, particularly, to consider agreed-upon, unified, and standardized approaches to its problems. These problems arise from a chaotic situation wherein everyone seems to be working independently. Much of the food composition data that exist are incomplete, incompatible, and inconsistent. William Rand (paper 2) presents some general ideas of food composition data as data, with emphasis on its variability

1

and its inherent incompleteness for solving specific problems. This paper points out some of the consequences of the current situation of incompatibility and inconsistency between data bases, with special attention to the lack of confidence in the whole field of food composition data. John Klensin (paper 3) presents the role of INFOODS as a "broker" of food composition data, helping to increase the accessibility and interchange of data, and co-ordinating the development of guidelines. A major point made by this paper is that the users should define the systems they need, and that computers should be viewed as tools to assist the user rather than as constraints.

The second group of papers is concerned with specific uses of food composition data. Ann Sorenson of the US Department of Health and Human Services and her colleagues at the University of Pittsburgh (paper 4) report on a number of international epidemiological research programmes, the problems they have with existing food composition data, and the implications for how food composition data should be organized and managed. Lenore Arab of the University of Heidelberg (paper 5) discusses her experience working in the area of nutritional epidemiology in Europe, focusing on practical problems and potential solutions that range from the difficulties of determining dietary intake to those of dealing with data from several different food tables, each with its own conventions. Ritva Butrum of the US National Cancer Institute (paper 6) presents many of the Institute's programmes, both within the United States and internationally, which require food composition data, and the problems that these programmes face without globally complete, consistent, and compatible data. Betty Peterkin, of the US Department of Agriculture (paper 7), shows how the data that USDA collects are used to satisfy various programmes in research, information, education, regulation, and food assistance, with special attention given to the assessment of the quality of the US diet and preparation of dietary guidelines. This paper lists differing aspects of food composition data which individual users require: sets of nutrients, various stages of processing or preparation, levels of statistical analyses, degrees of precision and accuracy, units of measure, and different formats. Carol Windham and her colleagues at Utah State University (paper 8) discuss the needs that university research in the areas of nutrition guidance and nutrition education place on food composition data. The paper also describes the components of and problems with the system that Utah State University and the Human Nutrition Information Center of USDA developed to fulfil these needs. The need for more and better data, and the importance of documentation to characterize the data, are stressed. Fred Steinke of the Ralston Purina Company (paper 9) discusses the uses that food companies make of food composition data, surveys the data that are currently being used, and maintains that safety and acceptability by the consumer are the primary concerns of the food industry.

The third group of papers is authored by developers, managers, and users of large food composition data bases. Loretta Hoover of the University of Missouri–Columbia (paper 10) discusses the general status of nutrition data bases within the United States in the context of the requirements and problems that the user faces in dealing with these systems. She introduces the concept of two tiers of users: those experienced users who are primarily interested in issues of accessibility, installation, updating, availability, and computational concerns as well as being aware of the problems of accuracy and applicability of different software and data bases; and those inexperienced users who are primarily interested in utility, compatibility, and cost. Any efforts to improve the status of food composition data need to be directed primarily toward the first tier of users; however, the results of improvement will shortly effect those in the second tier. Frank Hepburn of USDA (paper 11) describes the management of the USDA data base, illustrating how the data has evolved in response to a changing food supply and the more sophisticated needs of users. Marilyn Buzzard and Diane Feskanich of the Nutrition Co-ordinating Center at the University of Minnesota (paper 12) describe their system and the procedures that they have initiated to satisfy user needs. The specific needs stressed include: standardized methods of obtaining data; up-to-date and complete nutrient profiles; adequate documentation of data sources; quality control for the data base itself; and stability, comparability, and flexibility. Grace Petot of Case Western Reserve University (paper 13) discusses the management of a large academic system, cataloguing user needs and expectations.

The fourth group of papers is concerned with food composition data outside the United States. Hernando Flores of the Federal University of Pernambuco, Brazil (paper 14) illustrates the very real problems which arise from using the outdated INCAP tables to estimate consumption of important nutrients. Aree Valyasevi of Mahidol University, Thailand (paper 15) describes the status of food composition data in Asia, and reports on progress in the establishment of ASIAFOODS. Åke Bruce and Lena Bergstrom (paper 16) catalogue the data bases within Sweden and other Nordic countries, showing the profusion of data bases that can exist in a relatively homogeneous part of the world. Pamela Verdier of Health and Welfare Canada (paper 17) describes the national data base of Canada, how it was put together and managed, and how it is integrated into the nutritional activities of Canada.

The fifth and last group of papers discusses food composition data from several points of view slightly removed from the field. Joanne Holden and her colleagues at the Beltsville Nutrient Composition Laboratory of USDA (paper 18) discuss the problems of maintaining the quality of the data in data bases and of assessing and communicating that quality. This paper presents a specific system for evaluating published data on selenium, and discusses the aspects such a system must consider. George Beaton of the University of Toronto (paper 19) examines the effect of uncertainty in food composition on the estimation of nutrient intake from food consumption surveys, concluding that this variability is often less important than previously assumed. Reliability of food composition data is most important in those foods that make the greatest contribution to total nutrient intake, and the effect of improving reliability will be most evident when examining diets which include few foods. This paper also notes that these considerations focus on variability and that the presence of bias in the data is often an important source of error. Catherine Woteki of the US National Center for Health Statistics (paper 20) discusses in detail the US National Health and Nutrition Examination Surveys, which are a major user of the USDA data base. John Klensin of INFOODS (paper 21) examines food composition data bases from the point of view of modern information systems theory and practice, showing that the field has changed considerably since most data bases were first implemented.

2. THE USERS AND THEIR NEEDS

The Users of Food Composition Data

Data on the composition of foods are used by a wide selection of individuals and organizations, ranging from the Food and Agriculture Organization of the United

Nations, which calculates food balance sheets for countries around the world and advises them on the global allocation of foods and other resources, to individual shoppers reading labels in the supermarket in an attempt to plan the nutritional value of their next meal. The following list of examples illustrates the range and variety of users (see also the extended list in paper 13).

International Users (e.g. paper 15)

International agencies dealing with food aid are responsible for acquiring large quantities of food and transporting it to different locations in the world, often to meet the specific nutritional needs of populations. Here there is a strong need to know the constituents of the foodstuffs available around the world and, at the same time, the relevant nutrient situation within each country, in order to match the food supply to human physiological needs.

Governments need to know the composition of the foods that are imported in order to plan for satisfying the nutritional requirements of their populations and to protect these populations from contaminant and toxicologically active constituents.

Food industries that compete on the international market need to know the components of the local food with which they are competing, and also the nutrient content of potential ingredients for their products – ingredients which may come from widely separated parts of the world.

Epidemiologists have long realized that food is one of the most important components of an individual's environment. Studies of global disease patterns must be accompanied by data on global food and nutrient consumption patterns (see papers 4, 5, and 6).

National Users (e.g. papers 16 and 17)

National governments regulate what is produced within the country as well as what is imported in order to protect and guarantee the health of their populations [13]. Their activities include the setting of regulations and the monitoring of adherence to those regulations. They must also assess the nutritional status of their populations (papers 7 and 20); one important aspect of this is to determine what is being consumed and how this compares with established dietary requirements and allowances. All of these activities require the availability of reliable and extensive data on the composition of foods.

National programmes of feeding, such as in schools or in the military, require up-to-date information on the nutrient content of foods so that nutritionally adequate diets can be formulated.

University teachers and researchers require reliable and current food tables for a broad spectrum of activities that involve the relationship between health and disease and what people consume [2, 14] (papers 8 and 14).

Local Users

Below the national level, there are many activities concerned with the preparation and recommendation of healthy diets, in institutions such as hospitals and factories as well as in special segments of society such as the elderly.

Food industries regulate the quality of their foods by routinely analysing the components of their products, using available food tables. New formulations must adhere to nutritional and safety standards and food tables are used initially to select preparations for investigation (paper 9).

Individual Users

The diet and nutrition clinics which continue to increase in number in technically developed regions rely heavily on food tables for individual counselling. These clinics complement the activities of individual dietitians and physicians dealing with patients ranging from those with metabolic disorders to those who are over- and underweight.

Finally, the most extensive use of food composition data by the individual is, of course, the individual shopper scanning the ingredients list, nutrient contents, and percentages of requirements fulfilled on the labels of packaged foods.

Some Generalizations about Users

The extent and diversity of the users of food composition data are underlying problems of the field. These various users' groups not only make very different uses of the data, but also have different expectations and requirements. Thus it is important to identify the common threads that would permit the organization of users into groups who have similar needs and make similar demands on any food composition data system:

– System requirements are quite different for the user concerned with a limited set of foods that can be fairly well defined, such as those in a restricted geographic region or those suitable for a very specific diet, and for the user concerned with extensive, perhaps open-ended, sets of foods, with often a global distribution. The former group of users is usually able to define its needs and knows the questions it will ask of a data base, which can thus be embedded in a fairly rigid system, while the latter requires much more flexibility in its interactions with the data.

– The degree of precision or specific level of detail required divides users into a spectrum of groups, with some needing accurate data, some only approximate data or ranges, and others only an indication as to the presence or absence of components.

– Many users need to know the constituents of specific foods (e.g. diet evaluation) while others are interested in the foods that contain certain components (recipe or menu development). This distinction divides the users into two groups with quite different search and retrieval requirements.

– The distinction between whether the user is interested in a specific individual, for example for diet or menu planning, or in a group of individuals, for example to determine the nutritional intake of the population of a region, is essential in order to design a system to properly estimate various statistical quantities, such as confidence regions.

– Some users need only representative values for specific nutrients in specific foods. Others are interested in the extremes that a nutrient can attain (at a certain probability level), while others want to know how certain nutrients (and foods) correlate with one another. Each of these classes of users asks different questions, requires different amounts and types of data, and needs systems for usage which are superficially quite different.

5

The Needs of Food Composition Users

The basic needs of the users are: (a) high-quality data on foods and components, (b) ancillary data, and (c) facility of access and manipulation of these data to give the information desired.

High-quality Data

The most obvious need of the users is for data that represents the foods and components with which they are concerned. These data must be as precise as the particular usage requires, unambiguously identified, and easily accessible (see section 3 below).

Ancillary Data

For most users, the amount of a specific nutrient in a given specific food is not sufficient for their task (paper 2). Necessary ancillary data range from usual serving sizes to household consumption and individual nutrient requirements. The following list includes some of the major types of additional information that often may be needed:
- data describing the food composition data – where they came from, their quality, etc.;
- descriptions of representative serving portions, including weight/portion equivalents;
- brand names and identifications;
- recipes and ingredient lists;
- effects of processing and preparation;
- fortification rules;
- food/nutrient interactions;
- food/drug interactions;
- bioavailability;
- contaminants, toxicants, and allergenic compounds;
- dietary standards and nutrient requirements;
- food production and consumption;
- geographic distribution of disease.

Data Management

The mere existence of food composition and associated data is not sufficient for carrying out the tasks of most users. These data need to be acquired, merged, and manipulated in order to be turned into information. Thus the users require many of the tools of data-base management, with their use-specific tasks ranging from recipe calculations to grouping and imputation; these are discussed more fully in section 4 below. It is especially important to note that, with the ever-widening usages of food composition data, and with the increasing availability of those data, it is essential that there be consistency in the way these tasks are performed.

3. FOOD COMPOSITION DATA

There are three important aspects of food composition data themselves: (a) what data exist, (b) how good these data are, and (c) how easy these data are to obtain.

What Food Composition Data Exist?

The number of foods that have been analysed for their content varies tremendously around the world. Tables which include these data are available in a variety of forms (note that these tables are not entirely independent, since many of the basic data are shared among them):

- international tables (e.g. Platt's *Tables of Representative Values of Foods Commonly Used in Tropical Countries*);
- regional tables (e.g. FAO and USHEW's *Food Composition Tables for Use in Africa*);
- national tables (e.g. USDA's *Handbook No. 8*);
- food industry data bases (many major food companies have their own data bases);
- commercial data bases (there are a large number of diet analysis programs, which include data bases, available for purchase by individual consumers);
- local, special-purpose tables (many hospitals maintain data bases for menu planning and nutrition guidance);
- journal articles (a number of journals, such as the *Journal of the American Dietetic Association* and *Ecology of Food and Nutrition*, frequently publish papers containing food composition data).

While there is no complete index to all the food composition data that exist, several partial directories are available. The Food and Agriculture Organization (FAO) of the United Nations published a listing in 1975 which covered international, regional, and national tables [4]. This is now out of date and FAO has no plans for its updating. In 1986 INFOODS issued a similar directory of tables currently used [8]. EURO-FOODS (paper 5) and NORFOODS (paper 16) have prepared listings of data available within their regions. Within the United States, Loretta Hoover of the University of Missouri–Columbia annually issues a *Nutrient Data Bank Directory*, which includes characteristics and contents of currently available data bases [6]. Additionally, Darlene Hildebrandt, of the University of Washington in Seattle, issues a listing of *Computer Programs and Databases in the Field of Nutrition* [5].

With respect to the data that are available, in North Amerrica and Western Europe it is usually possible to find basic nutrient composition data for most common foods. However, there are many regions throughout the world where data on the composition of even the most frequently consumed foods do not seem to be available, or when available are seriously out of date (papers 14 and 15).

Beyond the problem of determining what data exist is the problem of determining what the available data represent. While often an introduction to printed tables will provide some indication of the analytic methods used, rarely is sufficient information given on how the food samples were gathered and analysed, and on how the data themselves were scrutinized and manipulated. Moreover, only infrequently is there any indication of the variability inherent either in the food or in the analytic method.

In terms of the data that are *not* available to potential users, no food composition data system contains values for all the components or foods desired by all users, and it is unlikely that any table or data base ever will, because of the rapid expansion of the number of foods and nutrients of interest.

Missing Components

The components of foods that are most frequently missing include:
– nutrients (especially trace minerals, some B vitamins, and lipid-soluble vitamins);
– subcomponents of nutrients and classes of nutrients, such as retinoids, carotinoids, fatty acids, starches, specific sugars, etc.;
– non-nutrients such as dietary fibres, xanthines, allergens, toxins, and selected contaminants;
– ingredients including additives.
While the situation is often that there are not good, reliable methods for assaying certain components [10, 12], many of the data do not exist simply because of the magnitude of the task of collecting them. Users with specific needs have two options: (a) they can generate the data themselves, gathering representative samples of the foods of interest and assaying them for the desired components, or (b) they can estimate (impute) the missing values from known data on similar foods and components. The first option requires resources that users rarely have available, while the second requires clear and well-defined rules for estimation, rules which do not currently exist. An associated problem is that rarely do data produced by individual users enter into the public store of food composition data, with or without appropriate documentation.

Missing Foods

While data on new, manufactured foods and foods infrequently consumed are often missing from food tables, there is a major and significant gap concerning the composition of "foods as consumed." Many of the data in tables represent foods that are "raw," but many (if not most) foods are eaten after being processed, stored, and/or prepared in various ways that may each affect at least some of the nutrient levels [1]. Moreover, often the data on prepared dishes in the tables are not the results of analyses but have been estimated by the compilers of the tables.

Two important types of "foods as consumed" are mixed dishes, such as stews and curries, and foods that are purchased already prepared, such as those obtained in a restaurant. For these foods, composition data, based on either analysis or estimation, must start from a recipe. However, it is often difficult to define, much less obtain, "standard" or representative recipes for most of these foods. If a recipe can be selected, one must then address the issues of labile or soluble nutrients, cooking losses, nutrient interactions, and fluid or fat loss (or gain) which can significantly alter nutrient concentration per unit weight. Additional problems arise from shifts in the availability and costs of ingredients which frequently lead to modifications of the recipes.

Efforts are proceeding in two general directions with respect to adding data to food tables. First, more analytic methods are being developed and analyses being conducted, and, second, discussions and research are being carried out to develop guidelines for making estimation more accurate. However, both these efforts must be greatly expanded, and co-ordinated, before users can devote their efforts to the using of food composition data rather than to the finding and completion of food composition data bases.

8

How Good Are the Data that Do Exist?

There is considerable variability in the quality of food composition data, and rarely is information about data quality available to the user (paper 18). It appears that the individual data that make up food tables and data bases have often undergone only limited scrutiny. While major tables choose their sources carefully and document these sources, this is expensive and time-consuming, and many data-base compilers do not give sufficient attention to this problem, leaving the responsibility of data quality to those from whom they acquire data. Similarly, estimation of data to fill in gaps in tables is frequently not performed with sufficient care, nor are these procedures documented, partly because of the lack of accepted guidelines. Clearly one conclusion that must be drawn from these considerations is that users must use food composition data cautiously.

Before considering how to improve the quality of food composition data, it must be pointed out that most variability in that data is not due to analytic error (papers 2 and 19). Of the number of factors that influence the observed levels of components in foods, it is true that several can be considered error, and their contributions to the overall variability of food composition data evaluated, categorized and, in some cases, minimized. For example, the analytical procedures introduce variability which can be minimized by following good laboratory techniques [11].

However, many other sources of nutrient variation are inherent in the foods themselves. These include geographical region of production, cultivar/species, changes in fortification levels, and agricultural practices in general [1]. Studies are needed to identify, characterize, and evaluate these several sources of variation to permit data compilers to provide users with food composition data that are less variable, perhaps through subdivisions of existing food categories – regional tables, dated values, and so forth.

The two general areas where major efforts are needed to improve the situation with regard to the quality of food composition data are: (a) improvement of the quality of the data per se, and (b) improvement of communications so that the user will be able to determine the quality of specific data of interest.

Improving Quality of Data

The long-range improvement of the quality of food composition data can best be achieved through amelioration of the measurement system (improvement of sampling techniques and analytic methods, development of standards for generating food composition data, development of training programmes in food analysis, and use of biological reference standards) and standardization of the procedures for manipulating data, including those for estimating data that are not directly available as analytic determinations. These are all essential efforts that will contribute to the reduction of the errors in food composition data.

Improving the Documentation of Quality of Data

Another area in which the field of food composition data demands a major effort is that of documenting more carefully the "context" of the data – those factors which can, and do, contribute to the variability of the data. This is essential so that users can be made aware of the potential problems of the data, and be given enough information to judge for themselves whether the data are of sufficient quality for their needs.

Additionally, there is a need for the development of an overall scheme to indicate the reliability of data (see paper 18 for a detailed discussion of this topic). For example, Exler [3] has described, and used, a procedure for evaluating existing data against fixed standards to produce a score, or "confidence code," for each data set. These confidence codes not only give the users of the data an indication of their reliability but also inform data generators where new data are needed, as well as providing data compilers with tools to rationally combine new food composition data with existing data.

Thus, food composition data are of uneven and often unknown quality, and users approach them with due recognition of this problem. Moreover, they must be aware of both the inherent variation in food composition data and the variation that can be introduced by the gathering and manipulating of data. Every effort must be made to make available to the user information concerning the food and its analysis that will provide insight into the reliability of the data and its suitability for a particular purpose.

How Accessible Are the Data?

The accessibility of food composition data is obviously essential to its usage, and has three key aspects: (a) finding the data (if they exist), (b) obtaining those data, and (c) determining the precise meaning of the data obtained.

Finding the Data

Determining whether the desired data exist and finding where they are located is discussed above and, as pointed out, represents a significant problem. Currently there is no complete, up-to-date, global catalogue of food composition data. It should be emphasized that the effort involved in compiling and keeping current such a catalogue, which includes enough information to be widely useful to the various user groups, is a major undertaking.

Obtaining the Data

The question of moving the data around – data interchange – is also a major problem because of the time and effort that must be devoted to the actual acquisition of food composition data – tables must be entered manually into the user's system, or programs custom-written to read specifically formatted tapes or disks. As the situation becomes more complex, with more generators, compilers, and users of data, the problems of data interchange will increase; and therefore dealing with them will consume more of the users' resources. Thus a major, essential task is the development of standardized guidelines for food composition data interchange.

Identifying the Data

Precisely identifying the data – determining exactly what food and what nutrient the numbers represent (the question of standardized terminology) – is key to the critical use of food composition data (paper 6). A standardized food-naming and classification system is critical to data entry, interchange, and retrieval, and currently no acceptable scheme exists. Although there are common elements that appear in the naming systems in most food composition tables, true compatability does not exist even among the most commonly used data sets. The development of a standardized global terminology for food composition data which addresses the associated problem of classification is an important task that needs to be initiated and accomplished as soon as possible.

4. A FOOD COMPOSITION DATA SYSTEM

Users obviously need more than just data – they need the machinery to interact with these data. This aspect of the subject can be discussed under the rubric "food data systems," used to describe the data and all the programmes or tasks involved in keeping the data relevant and available to the user.

The first point to be made is the distinction between those data bases and systems that are tailored for a single *specific purpose* or task, and those that attempt to be *general purpose*. This distinction is discussed at length by Hoover (paper 10) in terms of two tiers of users. It is important to note that the specific data bases are constructed from the general, and the validity of special-purpose data bases depends on the validity of both the data in the general data base and of the procedures by which they were selected.

The design and building of specific-purpose nutrient data bases and systems are straightforward since such systems can usually be completely defined in advance (although in fact they rarely are). Additionally, the actual data involved are usually fixed for the duration of the task. Much of the design effort here is focused on the user interface – making the system easy to use. A number of commercial firms supply such systems [5, 6]; however, a standard problem is that documentation of the source and quality of the data is frequently missing, leaving the user without guidance in this area.

General or broad-purpose nutrient data systems tend to focus on the data rather than on the details of interface with the user, although all systems must address this latter aspect. A number of papers given at the conference address the management issues (papers 7, 10, 12, and 13), others describe the magnitude and complexity of the task (papers 11, 15, and 17), and paper 21 focuses on the tools and concepts available to the system designer. The major points are summarized below, organized into the three categories of (a) the data themselves, (b) documentation of the data, and (c) preparation of subsets of the data.

The Data Themselves

The ideal general-purpose data base contains all the data that anyone might need, in a form that makes them readily accessible for any purpose. To approach this ideal it is necessary to be concerned with the following areas.

Data Acquisition

The data base must be updated continually and aggressively with new foods and new analyses, including re-analyses with better techniques, analyses of new products on the market, and new formulations of existing products. Thus standardized procedures must be implemented for routine collection of new nutrient information from available sources, including governmental publications, the scientific literature, and manufacturers' data on commercial products.

Data Consistency

An important aspect of adding data to a data base is that each new piece of data must be carefully evaluated for reliability. Moreover, all nutrient data files should be routinely checked for consistency, to identify possible anomalies and errors in the data. Such procedures could include, for example, comparing nutrient values within food groups or comparing actual data with predicted values. Thus the sum of the weight of the macro-nutrients plus ash and water theoretically should be 100 grams, while the sum of the caloric contributions of each macro-nutrient (including alcohol) can be compared with the total value for calories.

Having confidence in the individual data is one aspect of the question of the reliability of a data base. A check on the working of the entire data base, including a check on calculation procedures, can be provided by calculation of a selected, carefully constructed set of dietary records [7]. Such a test should be routinely carried out, with disagreement between successive runs carefully investigated and explained.

Documentation

General-purpose data bases need to contain information about the source and quality of each of their data points. At a minimum, the user should be able to trace back each piece of data – either to a source document or, in the case of analytic data, a laboratory reference; or, if it is estimated, it should be possible to ascertain just how this was done and from what other data. Moreover, it is important to maintain older data as part of the system. In the case of foods and food preparations which have been modified or are no longer on the market, data should be retained for comparison purposes, and so that dietary information collected in the past can be evaluated.

Preparation of Data for the Ultimate User

A major responsibility of the general-purpose data system is to prepare subsets of its data for the "front-line" users – these are the special-purpose data bases mentioned above. In order to do this at all well, such a system must support a flexible query language, an information data base that adequately describes the data, and sufficient manipulative machinery. Areas of specific importance are:

Access to the Data

The system should provide a variety of different ways to access the data. For example, foods should be indexed by food group and type of processing and preparation undergone, as well as by common name and food code number. Moreover, linkages to other data, such as food-specific quantity units, are also an essential part of retrieving the necessary data.

Aggregation of Data

Many users require data on quite general foods (for example, "apples" rather than "Red Delicious apples"). A general-purpose data base often contains some of these entries, with nutrient levels estimated by combining the data of several specific foods for which analytic data exist. It is essential that the data base include information on just how these estimations were calculated, and, further, that it provide the information, and perhaps the machinery, necessary for the users to make further combinations of data to suit their specific purposes.

Data Presentation

Presentation of data, either on a screen or in hard-copy reports, needs to be flexible to permit the design of special-purpose formats to meet specific user needs. For example, options for presentation of data should permit the display of calculated nutrients as a percentage of calories, or other calculated combinations of nutrient values, such as saturated fat as a percentage of total fat or in ratio to polyunsaturated fat. Other options might include comparison of calculated nutrient intakes with recommended standards for specific age–sex groups, or the reporting of nutrients for each individual food item, for single meals, for single days, or for the average of multiple days.

5. RECOMMENDATIONS

This conference reviewed the field of food composition data from the point of view of the user. A number of areas of concern were discussed and some specific issues raised concerning the development of the field. The conference was convened by INFOODS in part to gain insight into what INFOODS itself should be doing in the future. To this end it formulated a number of specific recommendations for INFOODS activities:

1. People working with the diverse aspects of food composition data are not strongly aware of the similarities of their efforts and of the issues they must deal with. This has led to a tradition of independent activities resulting in incompatibilities and duplication of effort. INFOODS is encouraged to work to develop a sense of community within the field. This effort, in part, involves communication, and it is therefore recommended that INFOODS publish a *Journal of Food Analysis and Composition* as well as compile international directories of food composition data and of workers in various facets of the field.
2. Standards or guidelines are needed in several areas:
   – data gathering: a manual detailing sampling and analytic methods for the gathering of food composition data;
   – terminology: comprehensive, international terminology for describing food composition data, especially the naming and classification of foods;
   – data interchange: a standardized scheme for the interchange of food composition data to facilitate the movement of such data around the world;
   – data manipulation: standardized statistical and mathematical procedures for manipulating data, especially in the areas of summary statistics and imputation of missing data;

– usage: suggestions on how food composition data should be utilized in various areas, such as epidemiology and dietary counselling.
3. Recently, attention has focused on the variability of human consumption and of human requirements for nutrients. This variability is complemented by the variability of food composition, an area which has been little studied and is poorly documented. This entire area of food data variability, reflecting inherent differences in foods as well as differences of analytic methodology, needs to be carefully studied, with special attention paid to identifying, measuring, and evaluating the components of variability, and additional attention to documenting and minimizing it where possible.
4. A major goal for the next few years is to make food composition data easily available on an international basis. It appears, however, that there may be legal difficulties developing. There are a number of consultants, companies, and even countries that produce and market data bases and food computer systems. These individuals and organizations are becoming aware of the commercial value of food composition data, and there are suggestions that the users of food composition data may soon have to deal with legal obstacles to the free interchange of their data. As such developments are monitored, these problems, and the users' options in response, need to be explored. A related problem, in the sense that it is a legal problem, is that of the responsibility for the accuracy and updating of data files. The question of who is legally responsible for errors that might result from calculations based on data bases is one that has arisen in other fields, and may well arise in the area of food composition data.

In summary, the participants at the INFOODS Users and Needs conference strongly supported the purpose and goals of INFOODS, offered the suggestions outlined above for what INFOODS should do, and urged INFOODS to begin working on them speedily.

REFERENCES

1. R. Bressani, "The Data Required for a Food Data System," *Food and Nutrition Bulletin*, 5(2): 69–76 (1983).
2. Å. Bruce and L. Bergstrom, "User Requirements for Data Bases and Applications in Nutrition Research," *Food and Nutrition Bulletin*, 5(2): 24–29 (1983).
3. J. Exler, *Iron Content of Food*, Home Economics Research Report, no. 45 (USDA, Human Nutrition Information Service, Washington, D.C., 1982).
4. FAO, *Food Composition Tables, Updated Annotated Bibliography* (FAO, Nutrition Policy and Programmes Service, Food Policy and Nutrition Division, Rome, 1975).
5. D. M. Hildebrandt, *Computer Programs and Databases in the Field of Nutrition. A Partial List*, 4th ed. (University of Washington, Academic Computing Center, Seattle, Wash., 1985).
6. L. W. Hoover, ed., *Nutrient Data Bank Directory*, 4th ed., Ninth Annual National Nutrient Data Bank Conference, Amherst, Mass., 18–20 June 1984 (Curators of University of Missouri, 1984; with supplement, 1985).
7. L. W. Hoover and B. P. Perloff, *Model for Review of Nutrient Database System Capabilities* (University of Missouri–Columbia Printing Services, Columbia, Mo., 1984).
8. International Network of Food Data Systems, *International Directory of Food Composition Tables*, 1st ed. (MIT, Cambridge, Mass., 1986).
9. W. M. Rand and V. R. Young, "Report of a Planning Conference concerning an International Network of Food Data Systems (INFOODS)," *A.J.C.N.*, 39: 144–151 (1984).
10. D. A. T. Southgate, "Availability of and Needs for Reliable Analytical Methods for the Assay of Foods," *Food and Nutrition Bulletin*, 5(2): 30–39 (1983).

11. D. A. T. Southgate, *Guidelines for the Preparation of National Tables of Food Composition* (Karger, Basel, 1974).
12. K. K. Stewart, "The State of Food Composition Data: An Overview with Some Suggestions," *Food and Nutrition Bulletin*, 5(2): 54–68 (1983).
13. J. E. Vanderveen and J. A. T. Pennington, "Use of Food Composition Data by Governments," *Food and Nutrition Bulletin*, 5(2): 40–45 (1983).
14. C. T. Windham, R. G. Hansen, and B. W. Wyse, "Uses of Nutrient Data Bases for Identifying Nutritional Relationships to Public Health and Nutrition Education in the United States," *Food and Nutrition Bulletin*, 5(2): 46–53 (1983).

Experiences with Food Composition Data:
The Context

1
INFOODS: Background and Current Status

VERNON R. YOUNG

Laboratory of Human Nutrition, Department of Applied Biological Sciences,
Massachusetts Institute of Technology, Cambridge, Massachusetts, USA

INTRODUCTION

Nutrient and food component data constitute a key and significant proportion of the body of knowledge required to understand the role played by the nutritional environment in human health and welfare. These data – their quality, quantity, and accessibility – constitute an essential resource for nutrition and food scientists and the various health professionals concerned with food supply and its implications for the well-being of individuals and populations at large. The data are generated and utilized by the international community of food and nutrition scientists, and thus it is essential that the continued development of this cumulative resource proceed in a way that promotes its maximum and effective exploitation for the resolution of nutrition and food-related problems both in this hemisphere and throughout the globe.

GROWING NEED FOR AND AVAILABILITY OF INFORMATION ON COMPONENTS OF FOODS

Information on the nutrient and non-nutrient composition of foods has been and continues to be used for a variety of purposes: assessment of the intake of and need for nutrient and non-nutrient constituents of foods by populations; meal planning and calculation of therapeutic diets; food production and nutrition policies; and the evaluation of food as a vehicle for transfer of toxic chemicals from the environment to man. To date, this information, particularly in relation to energy-yielding substrates and essential nutrients, has been obtained largely from food composition tables, a historical development of which has been presented by Hertzler and Hoover [5]. However, the increasing interest in and concern for the relationships between diet and food habits and degenerative diseases (including coronary heart disease, diabetes, hypertension, stroke, and cancers) has stimulated the further gathering of composition data on foods. Simultaneously, this has called attention to major limitations in the available data to support a variety of research activities in this area. These are especially evident in epidemiologic studies, where attempts are being made to characterize dietary differences between various populations exhibiting markedly different incidences of the diseases thought to be associated with diet and nutritional factors. Such studies have characterized inter- and intra-national differences in diet by broad categorizations, but suffer from the lack of detailed information on the nutrient and non-nutrient content of the enormous variety of foodstuffs consumed by different populations and subgroups within specific countries or geographic areas. Additionally, the introduction of nutrient labelling in the US, for example, has promoted the accumulation of data by the food industry on the composition of foods and their products. These data offer a further opportunity to expand the food data base.

Most of the information on the components of foods is still disseminated in the conventional, printed-page format; this is likely to continue as an important mode of communication for many users. However, there can be little doubt that the use of modern computer techniques will expand in the future, for a number of reasons:

1. Certain functions can be served more effectively through organization of a computer-based data bank than by conventional methods. Thus, one advantage of data dissemination in computer-readable form is the relative ease of updating; another is the possibility of carrying out sophisticated search strategies for material that has some specified set of desired properties, a problem that can be tedious or perhaps impractical when the information is tabulated in the conventional, hard-copy form.

2. The distribution of data in computer-readable form offers considerable advantages for its subsequent manipulation and usage.

3. The proliferation of small machines or of personal computers and interest by consumers in diet and food habits will inevitably result in an increased demand for computer-aided food composition data systems.

The number of food and nutrient data bases is increasing rapidly and many new entrepreneurs are becoming active. This is made evident, in part, by the growing number of meetings being devoted specifically to various problems associated with the development and uses of nutrient data bases [4]. Extensive nutrient data banks, such as the one established at USDA's Consumer Nutrition Center [6, 10], and at other locations throughout the world, illustrate a clear trend in converting existing food component data to a computer format.

With increased use of computer techniques, both for archiving food component data and for application of these data to various purposes, it becomes absolutely critical to examine carefully the ways in which food composition data are handled and used. Although computerization can be very efficient, it does not guarantee optimum or effective service to the final user of the data. A study by Danford [2] is illustrative of this problem. It was found in this study that when a diet based on common, unambiguous foods and simple serving sizes was submitted for analysis using 11 commonly used computerized data bases, considerable variation existed between the output for various nutrients. This is a particularly disturbing finding when it is recognized that all of the data bases had evolved from a common data source, namely USDA *Handbook No. 8*. A similar, but more detailed, comparative study of a dietary record was considered by Hoover [7], and she also found that the values for nutrients varied considerably between the various computerized nutrient analysis systems included in the study, even after elimination of coding judgements relative to portion size and selection of food items. Dwyer and Suitor [3] have also pointed out the uncertainty about the reliability and validity of the data in many computer-aided diet analysis systems.

Because of the problem of the reliability of the available data in nutrient data banks and the disservice to the scientific and technical community that can easily arise from non-standardized, non-evaluated data storage and dissemination, it may not be unreasonable to suggest that consideration be given to the registration of food component or nutrient data banks by an appropriate authoritative body. This is especially important, as has been emphasized by Lide [8], when attempts are made to provide data directly to the user, because both quality assurance and the opportunity for users to assess the suitability of the data for meeting a particular need are key factors.

GENERATION AND RECORDING OF FOOD COMPONENT DATA

In addition to the ways in which information on food components is or might be handled, issues related to the availability and validity of the original, analytical data are critical as we further consider the ultimate nature of INFOODS. It is probably unnecessary to point out that the generation and recording of food composition data represents a difficult problem in comparison with, for example, data in physics and chemistry, where information about atomic structure and bond energies is determined under highly standardized and reproduceable conditions. In the context of foods, analyses are made on living or deteriorating material, and there is also variability between and within foods. Although this is not surprising, since biological variability is one of the major prerequisites of life, it does add to the complex task of achieving an orderly accumulation of composition data on foods consumed by people. Furthermore, the enormous variety of foods eaten represents a crucial problem with respect to the identification of foods and their systematic classification.

The precision found in chemistry and physics rests on the ability to repeat experiments under highly defined conditions in different laboratories. Comparable precision in the determination of food composition is not achievable, but further efforts must be made in this area to improve the quality of food component data. Attempts must be made to develop suitable reference, certified food matrices to permit the establishment of rigorous quality-control procedures for the analysis of foods both within and among laboratories.

In any event, data should be generated and presented according to accepted criteria or guidelines, with adequate descriptions of the procedures utilized, including concern for sampling and coding, as well as analytical chemistry. In order for progress to be made, careful consideration needs to be given to developing recommendations and criteria for reporting experimental data on food components. In this way, some of the experience of the evaluators of food component data could be transferred to those who generate new information, leading potentially to an upgrading of measurement practices.

To enhance the process of organizing food composition data into an international and consistent ensemble, it is essential that the quality of the archival data be indicated by the inclusion of information on methodology and sampling procedures. As Touloukian [12] points out, if this is done the task facing the critical analyst will be made more feasible and the quality of the data available to the user will be increased.

INFOODS – AN INTERNATIONAL NETWORK OF FOOD DATA SYSTEMS: A FRAMEWORK FOR DISCUSSION

Some General Issues

Assuming, for the moment, that high-quality information on the nutrient and non-nutrient profile of foods is available, a data system might be depicted simply, for discussion, as follows:

$$Inputs \rightarrow Processing \rightarrow Outputs$$

With this scheme in mind it is necessary to consider the following tasks, as discussed in greater detail by Vickery [13]:
1. Definition of the objectives of the food data system.
2. Examination of the present status of collection, analysis, and interpretation of food component data.
3. Assessment of the general design of a system, including technical, economic, and operational feasibility.
4. Practical implementation of the design and evaluation of the system in terms of objectives.

These, as well as other tasks, can be listed and, indeed, discussed separately. However, the building of a functional international food data system requires an iterative procedure.

As indicated above, the first question is: "What should be designed?" The answer involves exploring with users their aims and needs. Thus, the nature of the information required must be carefully identified, and consideration given to the functions that this information is meant to perform. It is fundamental to know early in our programme of work what uses are intended for the output of food component data, for then it is possible to assess the usefulness of the current information and the format(s) in which it is obtained. After this, one can begin considering alternative and perhaps more effective ways of meeting the needs of users of food composition data.

It is worth emphasizing further that immediate output of any data system is a set of products, intended to provide the community with information believed to be of interest to them. Clearly, a prime aim in evaluating, analysing, designing, or perhaps improving systems of food data information is to determine what specific information is of interest and whether the existing or intended products do or will provide what is required. Again it is critical to know the users and the specific uses that are made or might be made of food composition information.

Desirability of an International Effort

The establishment of national and international standardizing organizations and data bases has done much to improve the intelligibility, accessibility, and transferability of scientific and technical data. Thus, the initiation of an international, co-operative effort in food component data systems is worthy of particular consideration, especially when reference is made to the importance of food in international trade, in national and international aid programmes, and in the broad area of international health.

The precedent for an international food component data system has already been set by the many instances of international, non-government organizations working in the field of scientific and technological data; examples are the programmes of CODATA (Committee on Data for Science and Technology), the International Council of Scientific Unions, international organizations concerned with problems of handling data, such as the Rome-based Intergovernmental Bureau for Informatics (IBI), and the United Nations and its specialized agencies.

National food and nutrient data services are being organized in increasing numbers. It would be a major advantage to the community of nutritionists, food scientists, and health professionals if these services were compatible and, indeed, readily accessible. There is thus a need to consider the co-ordination of regional, national, and international food data services. This might range from a centralized file with on-line access, at one end, to a looser co-operation based on common guidelines and standards for information gathering and handling, at the other end. These possibilities are at least technically achievable, since there are numerous computer systems capable of accessing large centralized data banks from remote terminals via telecommunications, as well as systems based on decentralized data banks, updated in interchange of defined format tapes.

The task of placing existing food composition data in machine-retrievable storage is substantial but can be accomplished [6, 10]. In terms of the human food supply, it is likely that this task will be done in piecemeal fashion even if a major commitment is made by the scientific community. As Abelson [1] has pointed out, if this problem is left entirely to individual initiative there will inevitably be enormous duplication of effort, and other problems will be neglected. Co-ordinating such an effort will require a high degree of international co-operation. However, international co-ordination is an attractive idea and we should strive to induce governments, academia, and food industries to provide support and sponsorship of these activities.

We must appreciate that private industries, governments, and non-profit institutions within and between countries will have their own distinct interests and responsibilities, and so a pluralistic approach to the generation, dissemination, and use of food component information is inevitable. It is, however, certain that, in view of the cost and effort involved, there are strong incentives for collaboration in building food component data bases, even if a multiplicity of information delivery systems turns out to be desirable. Furthermore, some central co-ordination is desirable, if only to encourage a degree of standardization and compatibility that will benefit all sources.

INFOODS at Present

Overall Construct

In consideration of the arguments presented above, and with the further assessments and recommendations of a group of experts convened to address the status of the field of food composition data [9], it was proposed to establish INFOODS, defined as:

An organization to promote international participation and co-operation in the acquisition and interchange of data on the nutrient composition of foods, beverages and their ingredients in forms appropriate to meet the needs of government agencies; nutrition scientists; health and agriculture professionals; policy-makers and planners; food producers, processors, and retailers; and consumers.

A cornerstone of INFOODS' work plan is the idea that a majority of people involved with food composition data can be categorized as carrying out one of three major activities: generating the data in the laboratory; collecting and organizing the data of others for use by others; or using the data in a number of ways. Of course, these three groups may overlap in that many individuals are involved with two or even all three of the activities. However, they are logically distinct, and it is helpful to consider them as categories for purposes of further discussion in this paper.

Thus, apart from the users and uses of food composition data, there are many other activities involved, which concern the generation and preparation of the data. These would include the analysis of food, the development of analyses, the compilation of specialized data bases, and the design of diet analysis programs. These activities are all ongoing now, but a problem that must be faced is that they are often being done independently. Furthermore they can be, and often are, done in different ways.

This problem becomes obvious when one considers the flow of food composition data: food → analysis → data base → user. Just as there are different users and uses, there are different specific paths one might take: ways of selecting foods, performing the analyses, combining replicate analyses, and organizing data bases – each might differ. These paths can produce different numbers; food composition tables or data bases therefore differ, and the numbers in the various tables may mean different things.

The independence of all these individuals and activities relating to food composition data have contributed to the fact that the state of food composition data today is simply not satisfactory. Problems do exist, and many of them can be organized around four specific issues: (a) data do not exist; (b) data cannot be accessed; (c) data are not complete; and (d) data are not accurate.

It was to attempt to resolve these problems that INFOODS was organized essentially as two networks: a network of food composition data, and a network of people interested in food composition data. In addition, a secretariat, with Dr W. M. Rand of MIT as executive secretary, has been organized to co-ordinate the various activities, with major financial support for this purpose coming from the US National Cancer Institute, and additional sponsorship from the National Heart, Lung and Blood Institute, the US Food and Drug Administration, and the US Department of Agriculture. Administrative support is received from the United Nations University to facilitate the international aspects of the programme.

The scope and precise nature of the networks referred to above continue to evolve but the conceptual outline of the system is shown in figure 1. Thus, it is anticipated that local centres or individual users obtain their data and assistance from regional centres which communicate with each other, with institutions, and with the INFOODS secretariat. These regional centres could be individual countries or groups of countries. Thus, if a user needed data about the nutrient consumption of an immigrant population, some local data might be available or a request made for additional data from a regional centre. It might be that a regional centre would not have all of the information needed, but would, in turn, request such from another region, perhaps via consultation with the INFOODS secretariat to learn whether other data did exist and where they might be found. Most important, this schematic outline emphasizes that the key to this endeavour is communication, often between quite different entities.

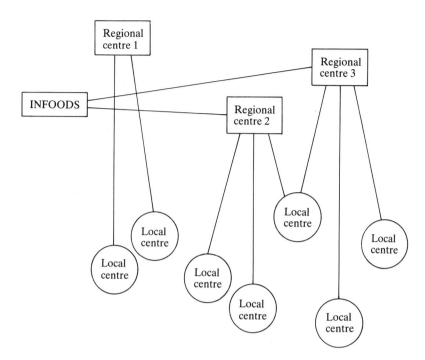

Fig. 1. An overview of the International Network of Food Data Systems (INFOODS).

The Approach Being Taken

To achieve a working network, linking together a number of disparate entities, there must be careful and detailed planning. In this context, figure 2 summarizes the major components of INFOODS. In order to achieve our desired results, three major working committees have been established to produce standards and guidelines.

The first of these is chaired by Professor A. S. Truswell of the University of Sydney, Australia. This is the terminology/nomenclature committee, and is concerned with the major problem that foods are not well specified, and have different names in various parts of the world. Often the same name is given to foods that may differ in origin and form. Any interchange of information needs an unambiguous definition of a food. It is of paramount importance that this work be carried out well and carefully. The tasks of the committee include: a review of existing systems, evaluating these systems in terms of their suitability for international exchange of food information, and recommendation of universal, international, standardized documentation language.

The quality of the food composition data themselves are the subject of the second working committee. This is chaired by Dr D. A. T. Southgate of the Food Research Institute in the UK. The overall responsibility of this committee is to develop guidelines and standards for the data used in the construction of food composition tables and electronic data banks. Its long-term goal is to improve the quality of data by standardizing methodologies, and, in the interim, to develop measures of quality and conventions

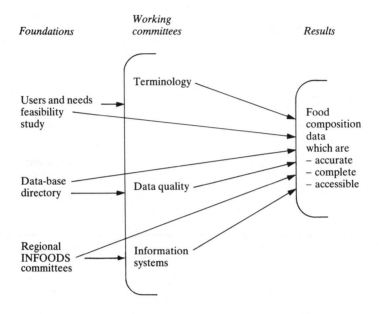

Foundations	Working committees	Results

Fig. 2. Schematic of approach being taken by INFOODS.

for data reporting. An initial product of this committee will be an updated and considerably expanded version of an earlier guideline for establishing food composition data tables [11].

The third committee, under the direction of Dr J. Klensin of MIT, is concerned with the information systems component of INFOODS. Several of the more important topics that this committee is undertaking are: (a) data interchange and communication standards; (b) ideal data set content; (c) data manipulation; (d) regional data centre design; and (e) local data centre design, including personal computers and small machines for individual use in office and field settings.

An essential activity that underlies the work of these three groups is a careful survey of the users and specific uses of food composition data. Information to be gathered includes: (a) the magnitude of usage and types of users; (b) the data most frequently needed; (c) the unmet needs and why they are not met; and (d) the users' desires and expectations. What is being sought is not just a catalogue of uses, but a careful and comprehensive examination of how they constrain the INFOODS network itself. For example, how quickly do most users need their data, what precision do they require, and what tools do they have to use the data? With detailed information of this kind it will be possible to arrive at a design for the forms in which the data will be delivered. It is these types of questions that are essential to the design of INFOODS.

Among other activities shown in figure 2 is the promotion of regional INFOODS groups. This is important to the network of people that is a fundamental component of regional INFOODS and essential to its overall success; the data network operates between people, and it is operated by and for people. One of the regional organizations,

EUROFOODS (Western Europe), began independently at about the time that INFOODS was established. Other regions include NORFOODS (Scandinavia), NOA-FOODS (North America), LATINFOODS (Latin America), and ASIAFOODS (Asia). Early plans are under way to promote regional groups covering the African continent; Australia, New Zealand, and the Pacific Islands; and also the Soviet and Eastern European nations. These regional groups are at different stages of development but are being organized with the intent of bringing together the people within a specific geographic area so that they can work efficiently on their own problems as well as interact more effectively within their region. Some of the activities each regional group might be or are involved with include: (a) regional data centres/clearing-houses; (b) identification of regional food composition data problems; (c) co-ordination of regional resources and activities; (d) contribution to development and review of implementation standards and guidelines; and (e) communication with other regional groups and with INFOODS.

In order to further the international contribution that our collective efforts might make toward resolving the many issues related to food composition data, the various activities of INFOODS are also being conducted with formal representation on the working committees by IUNS (International Union of Nutrition Scientists), IUFoST (International Union of Food Science and Technology), and IUPAC (International Union of Pure and Applied Chemists). Additionally, whenever possible the World Health Organization (WHO) and the Food and Agriculture Organization (FAO) are represented at the meetings in view of the interest in and concern for food composition data by these UN agencies.

SUMMARY AND CONCLUSIONS

Information on the components of foods embodies the heritage of our understanding of the role of nutrition in human health. It serves as a medium for the education of future scientists and as a principal reservoir of concepts and data to be applied to the resolution of nutrition and health problems. Some reasons have been given indicating that the need for reliable food component data will become more pressing. In parallel, the use of computer-based methods for the storage, processing, and retrieval of such data will accelerate. It is essential that, with such developments, a co-ordination of effort be undertaken at an international level to standardize the gathering, synthesis, and application of this information.

To meet this challenge we must assess, in detail, the users and uses of food composition data. From this assessment a series of design specifications for INFOODS will emerge. In this way it is hoped that the goal of effectively and efficiently increasing the quality, quantity, and accessibility of food composition data on a worldwide basis will be significantly closer to becoming a reality.

REFERENCES

1. P. H. Abelson, "Production and Use of Data in the Pure and Applied Sciences," in P. S. Glaser, ed., *Data for Science and Technology* (Pergamon Press, New York, 1981), pp. 5–9.
2. D. E. Danford, "Computer Applications to Medical Problems," *J. Paren. Enteral Nutr.*, 5: 441–446 (1981).

3. J. Dwyer and C. W. Suitor, "Caveat Emptor: Assessing Needs, Evaluating Computer Options," *J. Amer. Diet. Assoc.*, 84: 302–312 (1984).
4. H. Greenfield and R. B. H. Wills, "Tables of Food Composition: An Australian Perspective," *Food Tech. Australia*, 33(3): 101–130 (1981).
5. A. A. Hertzler and L. W. Hoover, "Development of Food Tables and Use with Computers," *J. Amer. Diet. Assoc.*, 70: 20–31 (1977).
6. F. N. Hepburn, "The USDA National Nutrient Data Bank," *A.J.C.N.*, 35: 1297–1301 (1982).
7. L. W. Hoover, "Computerized Nutrient Data Bases: I. Comparison of Nutrient Analysis Systems," *J. Amer. Diet. Assoc.*, 82: 501–505 (1983).
8. D. R. Lide, "Critical Data for Critical Needs," *Science*, 212: 1343–1349 (1981).
9. W. M. Rand and V. R. Young, "Report of a Planning Conference Concerning an International Network of Food Data Systems (INFOODS)," *A.J.C.N.*, 39: 144–151 (1984).
10. R. L. Rizek, B. P. Perloff, and L. P. Posati, "USDA's Nutrient Data Bank," *Food Tech. Australia*, 33: 112–114 (1981).
11. D. A. T. Southgate, *Guidelines for the Preparation of National Tables of Food Composition* (Karger, Basel, 1974).
12. Y. S. Touloukian, "Reference Data on Thermophysics," in B. Dreyfus, ed., *Generation, Compilation, Evaluation and Dissemination of Data for Science and Technology* (Pergamon Press, New York, 1975), pp. 138–146.
13. B. C. Vickery, *Information Systems* (Butterworths, London, 1973), p. 238.

2
DATA: The User Context

WILLIAM M. RAND

INFOODS Secretariat, Massachusetts Institute of Technology,
Cambridge, Massachusetts, USA

INTRODUCTION

It is the purpose of this conference to explore the linkage between food composition data and the user of these data. INFOODS has been set up to work towards "good" and "available" food composition data, and an essential first task is to establish just what is meant by "good" and "available." It is obvious that these terms only have meaning in the context of how the data are to be used, and thus a natural starting-point is to discuss and detail their users and uses. As a prelude to this, this paper first examines the context of food composition data usage, and then shows that the variability of the data is a key concept in unifying the various uses.

THE LINK BETWEEN THE USER AND THE DATA

Most users of good composition data have to balance three major, different types of information – answers to three different questions: what is in foods?; what do people eat?; and what do people need?

While these can be written as distinct questions they are closely interrelated. Discussion of any one of them without consideration of the others would be possible; however, it would be, for the most part, academic. Since this conference is directed towards the relationship between the user and the single question of what is in foods, it is important to realize that there must be at least implicit realization of the importance of the other two aspects – what people eat and what they need. The strict inseparability of these questions becomes more apparent if we examine more closely the relationships between foods and users, following just the two questions of what is in foods and what people eat. Figure 1 shows the four major components of this linkage: food, data, the individual, and the user. These are connected by four types of activity: analysis, data usage, consumption, and diet evaluation. It is obvious that, for the user to be able to do anything meaningful, the two paths from food to user must be both compatible and consistent.

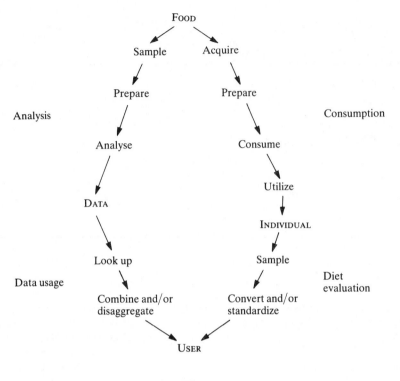

Fig. 1.

The linking of data and user, and how this link can be improved, is of special interest. It must be appreciated, however, just how constrained this link is by its following the analysis of the foods – for example, often we cannot get all the data we may think we need. Moreover, while we often feel that we really want to know precisely what is in foods, that information is really only useful if it can be correlated with what we can find out about what people really are eating.

Two additional points need to be made in conjunction with considering the data–user link in the context of our three questions. These are complementary and follow from the above. The first is that good, readily available food composition data, however we define good and readily available, will not solve all the users' problems. This leads to the second point that data are not of interest in and of themselves; they are of value only if they permit some generalizations to be made, some predictions or decisions.

THE VARIABILITY OF THE DATA

By the "goodness' of the data we mean suitability for the purpose at hand – how well the data will permit us to get on with whatever task we are involved with. For most users, finding a value in a table is sufficient. They will then use it as they need,

Table 1. Whole chicken egg, fresh, raw, per 100 g edible portion

Table	Water (%)	Energy (kcal)	Protein (g)	Fat (g)	CHO (g)	Calcium (mg)	Iron (mg)
USDA (new)	74.57	158	12.14	11.15	1.2	56	2.09
USDA (old)	73.7	163	12.9	11.5	0.9	54	2.3
United Kingdom	74.8	147	12.3	10.9	Trace	52	2
Federal Republic of Germany	74.1	167.08	12.9	11.2	0.7	56	2.1
Sweden	74.4	150	12.7	9.4	2.7	51	2.1
Denmark	74.6	155.8	12.1	11.2	1.2		2
Finland	74	145	12.8	11.7	0.5		2.5
Norway	75	155	13	11	0.7		2.1
Italy	73.9	156	13	11.1	1	50	2.5
East Asia	73.7	163	12.9	11.5	0.8	61	3.2
China 1	71	170	14.7	11.6	1.6	55	2.7
China 2	70.8	187	11.8	15	1.3	58	4.3
China 3	73	174	13.1	13.5			3.6
China 4	70	175	15.3	11.9	1.6	64	0
China 5	73	160	12.7	11.3	2	55	2.8
Republic of Korea	74	160	12.7	12.1	1.2		
Japan	70.7	199	12.2	15.2	0.9	65	1.8
Malaysia	73.2	166	13.3	12.5	0	57	3
India	73.7	173	13.3	13.3		60	2.1
Africa	77	140	11.8	9.6	0.6	45	2.6
Near East	72.8	160	12.1	11.4	1.2	55	2.9
INCAP	75.3	148	11.3	9.8	2.7	54	2.5
Brazil		163	12.9	11.5	0.8	61	3.7
Australia		160	12.6	11.6	0.8	54	2.4

a. Data are taken from standard national and regional tables.

assuming that that number is the best available estimate of some specific nutrient in some specific food.

If the user were so confused that he consulted an "expert," that expert might give either or both of two answers: (a) "the values in the different tables are really measurements of different objects, often by different methods," or (b) "the differences do not matter." While the first answer is probably true, the second is often false – in general it does matter. For example:

1. A person on a specific diet will receive differing advice depending on what data base is used to analyse his/her food intake.
2. A small difference in an individual diet can become a large difference when projected to a population estimate – the level at which important decisions such as resource allocation are made (table 1).
3. In general, apparent and unexplained inconsistencies reduce the confidence in all such data, in the system which provides such data, and in the science that works with such data.

The problem hinges on the fact that it is unlikely that these data really are inconsistent – they only look inconsistent. Most data bases present a single value for a specific

Table 2. Calcium (mg) in 100 g milk

USDA (8.1)	Mean = 119	SE = 0.251; N = 1,054 (SD = 8.15)
McCance and Widdowson	Mean = 120	Range = (110–130)
Souci/Fachmann/Kraut	Average = 120	Variation = (107–133)
Swedish NFA	Average = 113	Variation = (100–122.1)

nutrient in a specific food. This leads back to the first answer above – that different things were measured. If we look closely at the "food" component of figure 1 we note that it should be expanded as in figure 2, to show that every sample of a food is quite likely to differ from every other sample, and this is before the chemists take over and add their own variability.

While it is important to realize that there are a number of specific sources of food composition data variability, the major point is that few tables even hint that such variabilities exist. Moreover, those tables that do, such as those shown in table 2, do not do so in a consistent fashion, nor are they very helpful about just how to use this added information.

The point to be stressed is that there is not a single food of each kind – there is no Platonic ideal "egg." Foods are not mathematical ideals, but must be considered as probabilistic or statistical objects, with statistical distributions of their nutrients. Any compilation or use of food composition data must be firmly based in this fact.

We are faced with the fact that a data base needs to contain more than just a single value for each food-nutrient combination. The description of a distribution is not straightforward; few distributions can be described adequately with just a few numbers. (The obvious counter-example is the Normal or Guassian distribution, the familiar bell-shaped curve. However, few measurements follow this distribution precisely.) The "statistics" that can be used to describe an arbitrary distribution include

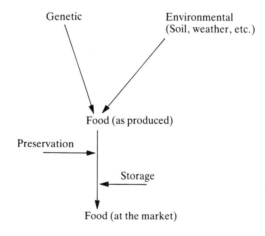

Fig. 2.

the mean, mode, median, quartiles, percentiles, standard deviations, and mean deviations. Each has its adherents and rationals.

The improvement of food composition data requires careful investigation of both where the data come from and what they are to be used for. Inherent in the viewing of food composition data as "data" are several implications that need to be stressed.

1. Each type of user is likely to require different statistics. These need to be carefully defined and justified. For example, someone wanting to estimate intake would perhaps be satisfied with a mean or median value, while someone worrying about meeting requirements would want an upper or lower limit. In order for data banks to be well designed, for them to include "good" data, each user must decide the best data representation for his/her specific application.
2. Data banks must be designed to provide information about the variabilities of their holdings. Ultimately this requires access to raw data, but, well before that, standardized and documented algorithms for data manipulation are needed.
3. Users must be made aware of the inherent variabilities of the data, of the magnitudes and implications of these variabilities, and of the procedures for handling this inescapable aspect of food composition data.
4. The sources of data variabilities need to be sorted out as a preliminary step, estimating their magnitude, exploring their importance, and reducing those that can be reduced by approaches ranging from standardizing analytic techniques to developing regional values.

3
The INFOODS System

JOHN C. KLENSIN

Laboratory of Architecture and Planning and INFOODS Secretariat,
Massachusetts Institute of Technology, Cambridge, Massachusetts, USA

INTRODUCTION

A major goal of INFOODS is to improve the availability and accessibility of food composition data. This goal, coupled with the existing structure within the field and the resources currently and likely to be available to INFOODS, mandates that the basic structure of INFOODS consist of a small secretariat, multiple regional centres, and guidelines about how "things" should fit together and operate. This paper presents an overview of how we expect INFOODS to operate as a nutritional "system," with special emphasis on data interchange from the perspective of the user.

DATA INTERCHANGE AND REGIONAL CENTRES

INFOODS is basically a collaboration of people and organizations involved with the generation, collaboration, and use of food composition data. Key to this is the concept of moving data around – data interchange. The current model of the operation of INFOODS shows a number of regional centres, each acting as a focus for composition data within its region. We hope and expect that there will not be more than about a dozen such regions, and that industries and even governmental data banks will not constitute regions. Within a region, user centres and user groups – ranging in size from individual desk-top computers to large governmental or university installations – will use the regional centre as their access point for data from other regions, and perhaps as a means of accessing data from within the region. No particular model of regional centre operation is required by this. We expect that some regions will maintain all regional data centrally, that others will maintain no data but will have convenient mechanisms for accessing data from distributed repositories, and that still others will operate with a mix of local and distributed data. We see the flow of transactions as:

1. User facility queries the regional centre as to whether it has particular data available and how it is obtained.

2. If the regional centre does not have the data, and is not aware of its availability within the region (or whether another regional centre has it), it makes an INFOODS inquiry as to where the data might be found. In order to make this inquiry the only requirement is that there be a single international focus once the regional centres are in full operation. It is an inquiry about the location of data ("if this exists, who has it?") rather than about the data, and does not necessarily imply any computerized facility or electronic communication at all.

3. INFOODS either responds to the original regional centre with the availability information, in which case that regional centre requests the data from a second centre, or it requests that the second regional centre forward the data to the first.

4. The second regional centre does whatever is needed to obtain and prepare the data within its region, then forwards that data (in INFOODS interchange format) to the first regional centre. The interchange format is insensitive to mode of transmission – electronic networks, magnetic tapes, or even messages written on pieces of paper. Transmission modes will be worked out between pairs of regional centres on the basis of what is available and the real or anticipated level of demand.

5. The regional centre forwards the data to the user facility, possibly after some format conversion or editing operations agreed to within the region.

REGIONAL DECISIONS

This approach implies several decisions that each region will make separately. While INFOODS may make suggestions in these areas, the effective operation of the network will not depend on those suggestions being followed (or even asked for and supplied).

The first of these decisions is where the data themselves will be retained. The overall INFOODS notion requires only that a regional centre be able to provide requested information when it is requested, or within some reasonable time thereafter. The data could all be kept at a regional facility, or some could be kept centrally and some at various sites within the region, or all could be kept at local sites, with the regional centre acting as a collection and redistribution point only. Indeed, it would be possible for a region not to have any centralized computing facilities at all, but simply to receive requests for data and dispatch those requests to facilities within the region that had the data and could respond to the requests.

If we consider the possibility of each local facility – a country, a ministry, or academic or research facility – having its own unique computerized tools for managing nutrient data, the importance of the standard format for the interchange of data becomes even clearer. The INFOODS notion is that data moved between regional centres will be organized into this interchange format, with translation to and from that format occurring within the regions. The second regional decision will concern just where that translation is made. Will it be made at the regional centre, which might maintain a table of the formats required by local facilities within the region? Or will local facilities be expected to accept (and, if they are producers, create) data in the interchange format? We expect that this decision, too, will differ from region to region. It may even differ within a region, with some local formats being supported by the regional facility and local facilities that require other formats being required to translate from the interchange form themselves.

Third, a region will have to make sure its own decisions about how much data originating outside the region it should retain and for how long. At one extreme, the nature of the interchange and query arrangements should be such that there would be no technical obstruction to a single region assembling all of the world's data and retaining them locally. Since the cost of doing so includes an ongoing effort to keep that body of data current (or to determine when subsets of the data are no longer current), we have recommended that no region actually attempt to do this. A regional facility that discovered that particular data were requested repeatedly might reasonably decide to retain them, rather than requesting them from another region each time an inquiry arrived. At the other extreme, a region that chose to have no regional centre at all, but only a communications network, would presumably retain no data from outside the region except at local facilities.

LOCAL DECISIONS

Just as there are decisions that we expect each region will make separately, even if several of them reach the same conclusions, there are decisions that we expect will usually be made at facilities within the regions. For convenience, we describe these as "local decisions" and "local facilities," recognizing that some facilities that operate within regions and below the regional level may, themselves, operate as sub-regions with several levels of smaller or more local machines dependent on them for data or other services.

Each local site will be able to make its own decisions about what programs it should run and what hardware they should be run on. That flexibility is needed for at least two reasons – the need to be able to accommodate local interests or requirements for different types of data processing or entry, and the reality of constrained choices in hardware, and sometimes software, in various parts of the world (for example, if an institution has decided to run only IBM mainframe hardware and no stand-alone machines, then there is no value in a recommendation for "standard" software that requires a desk-top machine).

Not unlike the regions, each local facility will have to make its own decisions about what data are retained and for how long. While at the regional level this question applies primarily to the foods to be retained, at the local level there is the additional question of what data – nutrients, quality information, data-source information, other descriptive information, non-nutrients, and so forth – about each food should be kept as well.

Similarly, requests for data to be imported into a particular local facility will need to be made at that facility. Neither INFOODS nor a regional centre can make sensible recommendations about what data (especially extra-regional data) a particular local facility should retain or make available to its users. In this context, any software, especially end-user (small machine) and the user interfaces of regional software maintained by INFOODS, will be produced primarily as demonstrations of the feasibility of the general structures, of at least one way to do things, and, in the case of interchange formats, references as to how those formats are handled. INFOODS is also likely to produce some subroutine-level codes that will be suitable for embedding in other systems to provide debugged interfaces to facilities defined as part of the INFOODS work. There is no obligation on any facility, local or regional, to utilize any of that software directly, and its principal purpose will be demonstration and reference. None the less, we expect that it will be directly useful to some facilities, and they will certainly be invited to use it if they wish. The decision to do so will be another decision that can be made regionally or locally, as appropriate.

If we have a single group of recommendations for everyone, whether local facility or regional centre, it is that programs and systems be designed on the assumption that change, probably a great deal of change, will occur and is inevitable. In particular, as technical groups and individual scientists in different parts of the world progress with their work, we can expect to see explosions in both data and requirements: more foods, more nutrients, more desire to accommodate non-nutrients and source, quality, and description information, and a continuing flow of new and updated data as analytic methods improve. For those of us who have become attached to relatively small machines, there may also be changes in systems and architectures every few years as our perceived needs expand to match new generations of hardware and software. We argue strenuously below that the best system may be one that is closely tailored to a particular set of needs and users, with no pretence at being "general purpose," but even such special systems should be designed so that the next study to be performed, or the next nutrient to be entered, does not completely disrupt them.

"What does the user need?" One theme that surfaces frequently in discussions is that users have only two important needs, "data and more data." We suggest that this is not true. The reason why it is not leads us to the first of our major system design suggestions. Data, as data, are essentially worthless. Computer systems should serve us in two important ways – to organize and catalogue data so that they can be managed, retrieved, and processed, and to provide assistance to process the data into *information*, something that we do need and probably need more of. Information has to be defined in terms of particular needs, and it is essential that the needs of food composition data be articulated and documented.

The Uses of Food Composition Data

4
Need for a Standardized Nutrient Data Base in Epidemiologic Studies

ANN SORENSON

Office of Assistant Secretary of Health, Office of Disease Prevention and Health Promotion, Department of Health and Human Services, Washington, D.C., USA

HYUN KYUNG MOON LEE and MARGARET F. GLONINGER

Department of Epidemiology, Graduate School of Public Health, University of Pittsburgh, Pittsburgh, Pennsylvania, USA

INTRODUCTION

The current concern in the area of nutrition, diet, and chronic diseases such as coronary heart disease, diabetes, hypertension, stroke, and cancer has stimulated an interest in detailed chemical data on foods, and subsequently called attention to some major deficiencies in the nutrient data bases available to support a variety of research activities in this field. This has become especially evident in epidemiologic studies charting the dietary differences between various populations which have markedly different incidences of chronic diseases thought to be associated with diet. Such studies have been able to show international differences in diet by broad, nutrient-food categorizations, but they are limited in assessing dietary risks because of a dearth of detailed information on the nutrient content of many of the foods consumed.

Current diet and disease studies require data on the human requirements or allowances for essential nutrients and quantified data on the ability of the food supply to provide these nutrients. In addition, other components of foodstuffs, including contaminants, intrinsic and extrinsic toxicants, and non-nutritive chemicals, should be identified and quantified to elucidate possible etiological relationships between diet and major public health problems.

Ideally diet and disease studies should take into account the synergism and inhibitory factors of nutrients with each other and with other environmental factors. Factors relating to bioavailability could be calculated and mathematical algorithms developed to adjust intake for other conversion factors related to gut metabolism. For example, the conversion factors for enhanced absorption of non-haem iron in the presence of ascorbic acid can be stored as part of the data system.

At present, no food composition data system exists that provides complete and systemic nutrient and non-nutrient information on food composition. Many foods commonly included in research studies have not been assayed. There are no values for some nutrients in some foods, and in other cases the existing food composition analyses are inadequate. Much of the problem stems from the complex and dynamic nature of human food supplies and the lack of reliable analytical chemical techniques for determining food composition for some food constituents.

LIMITATIONS OF DIET-RELATED EPIDEMIOLOGIC STUDIES

Suitable and up-to-date food composition tables are practical tools for the identification of dietary problems and the planning of intervention programmes. Epidemiologic studies are largely dependent on food composition data bases because of the cost and impracticability of obtaining and assaying foods from the large number of free-living subjects required for such studies. Therefore, food composition data bases should, whenever possible, give reliable representative data for indigenous foods reflecting the effect of growing conditions and treatment before consumption. They should include a wide variety of nutrients, making possible a comprehensive study of nutrient intake.

Advances in analytical chemical technology and the advent of high-speed computers have made feasible the processing of complex human diets. However, there is substantial criticism of diet-related epidemiologic research because the results of many studies have been weak, inconclusive, or equivocal, and at variance with animal models and *in vitro* evidence. Many problems with population-based diet studies relate to the following issues: (a) determining the strength of diet relationships to disease states which have multiple histologic and physiological characteristics; (b) identifying the significant dietary causal risk factors affecting the disease state; (c) having an incomplete or inappropriate nutrient data base to analyse data; (d) conducting studies with weak designs and limited technology; and (e) making inappropriate comparisons between study variables. Epidemiologic research related to diet and colon cancer can be used to illustrate how some of these problems can be influenced by food composition data, which in turn can influence the outcome of such studies. Colon cancer was selected as the example because it is a disease that has been strongly implicated with diet.

FACTORS INFLUENCING DIET-RELATED EPIDEMIOLOGIC STUDIES, USING DIET AND COLON CANCER STUDIES AS AN ILLUSTRATION

Searching on the key words "diet or dietary" and "colon cancer," "colonic neoplasms," or "sigmoid neoplasms," a MEDLINE literature search yielded 166 citations dating back to 1980. Twenty-six or 16 per cent of these studies were population-based or epidemiologic in nature. Thirty-three population-based studies reported after 1977 were identified by cross-referencing colon cancer with dietary risk factors. These studies have been summarized in table 1. The studies have been grouped according to the most commonly cited dietary risk/protective factors: dietary fibre, fat/meat, beer/alcohol, and cruciferous vegetables. The headings in table 1 list major components of epidemiological studies, each of which can effect the outcome of the study. The major types of study design as seen in the table are: ecological and food disappearance studies, retrospective (case-control) studies, cross-sectional surveys, and prospective (cohort) studies. In addition to choosing the appropriate study design, the investigator must also decide how to collect dietary information.

Though there are many variations of each, there are four basic dietary data collection tools: diet diaries, diet recalls, diet histories, and food frequencies. If data on specific food or food-group intake or availability is obtained for individuals or groups, the information can be transformed into nutrient intake by interfacing the food intake data of study respondents with a food composition data base.

Each technique has inherent strengths and weaknesses. Retrospective data collection methods are subject to respondent memory bias while diary methods tend to distort usual intake patterns. In addition these standard methods measure different aspects of dietary intake. Therefore there will be differences in study outcome depending on the food-intake data-collection instrument chosen. (Notice that all four intake tools were employed in the studies reported in table1.)

The type of food or nutrient data base selected is dependent on the study design, the data collection method, the study objectives, and the endpoints to be measured. However, a lack of standardized definitions of dietary study variables has been a major weakness in interpreting study outcomes. Definition has presented problems for developing standardized food names as well as for food composition tables. For example, dietary fibre, the first risk factor listed in table 1, is a complex of a number of physically and chemically different entities found in foods. They include cellulose, hemicellulose, lignins, pectins, and gums, and the ratio of these materials varies in fibre-containing foods. Until recently, data bases reported only crude fibre values, in which food samples were subjected to strong acid and then alkali solutions. These values are not equivalent to dietary fibre, which is the residue of undigested food.

The last column in the table describes the outcome or risk-factor association found in the studies. Drawing correct conclusions from the data concerning the strength of association of study variables and the attributable risk for diseases is dependent on choosing appropriate statistical tests. In addition one must control for confounding variables and adjust for co-variables. Unlike other clinical or laboratory studies, epidemiological studies are based mainly on relative rather than absolute differences of risk factors between exposed and unexposed groups. However, these studies lose power if real differences exist in the nutrient content of foods consumed by different population groups. This problem is analogous to regressing to the mean by not utilizing significant differences in food composition consumed by study populations. Increasing the power of a study is important, since the influence of diet is often obscured by stronger overriding etiological factors encountered in multi-etiological chronic disease studies. Also, epidemiologic methods and techniques are sometimes inadequate or inappropriate for the evaluation of diet and disease relationships, especially if one assumes that nutrient variables are independent of other dietary or environmental factors. Furthermore, much of the confusion in outcomes of diet-related epidemiologic research may stem from inappropriately comparing studies that differ in design, analytical techniques, or food composition data bases.

SOME POTENTIAL PROBLEMS WITH INCOMPLETE AND NON-STANDARDIZED NUTRIENT DATA BASES

Food composition data is required on many different levels of complexity and in forms that are readily computerized. In addition, users of such nutrient information need computerized composition data in a variety of formats that are not readily compatible

Table 1. Selected epidemiologic studies of dietary factors and colon cancer reported after 1977

Ref no.	Year published	Place of study	Type of study	Study method	Source of data base	Risk factor association[a]
Dietary fibre						
12	1977	Finland, Denmark	Cross-sectional	Dietary record	D10	+
23	1977	India	Cross-sectional	Food frequency	−	+
5	1978	USA	Case-control	Food frequency	−	+
21	1978	Utah, USA	Ecological study	Food consumption	−	0
22	1978	Scandinavia	Cross-sectional	Dietary record	D7	+
37	1978	South Africa	Ecological study	−	−	+
20	1979	Many countries	Ecological study	Food consumption	D11	0
27	1979	4-country	Ecological study	Food consumption	D11	+
1	1979	UK	Ecological study	Food consumption	D4	0
15	1979	Scandinavia	Prospective study	Food frequency	D1	0
29	1979	Israel	Case-control	Food frequency	−	+
13	1979	Canada	Case-control	Diet History	D7	0
3	1980	Kenya	Prospective study	Food frequency	−	+
26	1980	Australia	Migrant study	Food consumption	−	+
35	1981	Israel	Case-control	Food frequency	D1, D2, D3	+
16	1982	Denmark, Finland	Cross-sectional	Dietary record	D4, D5, D6	+
6	1982	4-country	Cross-sectional	Dietary record	D4, D5, D5	+
28	1983	Canada	Case-control	Food frequency	D7	+
33	1983	3-country	Cross-sectional	Diet history	D5, D9	+
31	1984	7-country	Ecological study	Food consumption	D11	+
19	1984	SDA in USA	Ecological study	Dietary record	D7	+
Fat and/or meat						
12	1977	Finland, Denmark	Cross-sectional	Dietary record	D10	−
5	1978	USA	Case-control	Food frequency	−	−

44

Table 1 (*continued*)

Ref no.	Year published	Place of study	Type of study	Study method	Source of data base	Risk factor association[a]
40	1978	Buffalo, USA	Case-control	Food frequency	–	0
21	1978	Utah, USA	Ecological study	Food consumption	–	0
22	1978	Scandinavia	Cross-sectional	Dietary record	–	–
27	1979	4-country	Ecological study	Food consumption	D11	0
1	1979	UK	Ecological study	Food consumption	D4	–
20	1979	Many countries	Ecological study	Food consumption	D11	0
40	1979	Many countries	Ecological study	Food consumption	D11	–
9	1980	Japan	Case-control	Diet history	–	0
13	1980	Canada	Case-control	Diet history	D7	–
26	1980	Australia	Migrant study	Food consumption	–	–
32	1980	USA	Ecological study	Food consumption	–	–
35	1981	Israel	Case-control	Food frequency	D1, D2, D3	–
18	1981	Hawaii, USA	Cross-sectional	Food frequency	D7	0
16	1982	Denmark, Finland	Cross-sectional	Dietary record	D4, D5, D6	0
17	1982	UK	Ecological study	Food consumption	–	0
24	1983	Greece	Case-control	Food frequency	–	–
28	1983	Canada	Case-control	Food frequency	D7	–
4	1984	USA	Cross-sectional	Dietary record	D7	–
36	1984	Hawaii, USA	Prospective study	24-hour recall	–	–

Beer and/or alcohol

Ref no.	Year published	Place of study	Type of study	Study method	Source of data base	Risk factor association[a]
7	1977	USA	Ecological study	Food consumption	–	–
22	1978	Scandinavia	Ecological study	Dietary record	–	–
27	1979	4-country	Ecological study	Food consumption	–	–
1	1979	UK	Ecological study	Food consumption	D4	0
15	1979	Scandinavia	Prospective study	Food frequency	D10	0

Table 1 (*continued*)

Ref no.	Year pub- lished	Place of study	Type of study	Study method	Source of data base	Risk factor association[a]
11	1980	Hawaii, USA	Cross- sectional	Food frequency	–	0
24	1983	Greece	Case- control	Food frequency	–	0
28	1983	Canada	Case- control	Food frequency	D7	0
14	1983	SDA in Denmark	Prospective study	Food consumption	–	–
30	1984	Nebraska, USA	Case- control	Diet history	D7, D8	–

Cruciferous vegetables

8	1978	Buffalo, USA	Case- Control	Food frequency	–	+
9	1980	Japan	Case- control	Diet history	–	+
24	1983	Greece	Case- control	Food frequency	–	+
28	1983	Canada	Case- control	Food frequency	D7	0

a. Risk factor: "–" = harmful effect; "0" = no effect; "+" = protective effect.
Sources of data bases:

D1. Y. Guggenheim, N. Kaufman, and A. Reshaf, *Food Composition Tables* (Ministries of Health and Culture, Government School of Home Economics and Nutrition, Romema, Jerusalem, 1980).
D2. R. M. Freely, P. E. Criner, and B. K. Watt, "Cholesterol Content of Foods," *J. Am. Diet. Assoc.*, 61: 134–148 (1972).
D3. R. M. Narayana and M. Polacchi, *Food Composition Table for Use in East Asia*, part 2 (NIAMDD; NIH, DHWS, Bethesda, Md., 1972), pp. 298–301.
D4. D. A. T. Southgate, "Dietary Fibre: Analysis and Food Sources," *A.J.C.N.*, Suppl. 31: s107–s110 (1978).
D5. R. A. McCance and E. M. Widdowson, eds., *The Composition of Foods*, 4th ed. (HMSO, London, 1978).
D6. W. P. T. James and O. Theander, eds., *Analysis of Dietary Fiber in Foods* (Marcel Dekker, New York, 1981).
D7. US Department of Agriculture, "Composition of Foods: Raw, Processed, Prepared," *Agriculture Hand- book No. 8* (Science and Education Administration, USDA, Washington, D.C., 1968; expansion, 1972).
D8. US Department of Agriculture, *Nutritive Value of American Foods in Common Units*, Agriculture Hand- book No. 456 (US Government Printing Office, Washington, D.C., 1975).
D9. Consumer and Food Economics Institute, *Nutrition Value of Foods* (USDA, Washington, D.C., 1971).
D10. Laboratory analysis.
D11. FAO, *Food Balance Sheets* (FAO, Rome, 1977, 1980).

with currently available data tapes. Thus, many epidemiologists are developing special- ized data bases, usually by modifying or adding to the USDA tapes.

For example, almost every researcher who begins nutrition-related clinical or popula- tion-based studies begins by finding and purchasing a data base that must then be modified (usually by a review of the literature) for the specific foods or nutrients under study. However, the uncoordinated creation of such data bases makes it virtually impossible to compare nutritional studies that utilize different data bases even when

Table 2. Selected nutrient composition of several varieties of cabbage reported by two different food composition tables (amount per 100 g)

	Kcal	Fibre (g)	Fat (g)	Vit. A (IU)	Vit. C (mg)
Bowes and Church					
Headed	24	0.8	0.2	130	57
Red	31	1.0	0.2	40	61
Savoy	24	4.6	0.2	200	55
Chinese	14	0.6	0.1	150	25
Spoon	16	0.6	0.2	3,100	25
USDA					
Headed	24	0.8	0.18	126	47
Red	27	1.0	0.26	40	57
Savoy	27	0.8	1.0	1,000	31
Chinese	13	0.6	0.20	3,000	45
Spoon	16	0.6	0.20	1,200	27

Sources: A. de P. Bowes and C. F. Church, eds., *Food Value of Portion Commonly Used*, 12th ed., rev. C. F. and H. N. Church (J. B. Lippincott, Philadelphia, Pa., 1975); US Department of Agriculture, "Composition of Foods: Raw, Processed, Prepared," *Agricultural Handbook No. 8–11* (Science and Education Administration, USDA, Washington, D.C., 1984).

those data bases are relatively well known and documented. In addition, the repeated modification of existing USDA tapes duplicates effort and increases costs that could be minimized by having an available standardized data system.

With reference to the colon cancer literature, ten different nutrient data base sources were cited in the studies listed in table 1. The variability of data can be demonstrated by the differences in data on the nutrients contained in foods in even well-known food composition tables, as shown in table 2. Here, reported vitamin A levels show differences between source A and B for savoy, Chinese, and spoon (*pi-tsai*) cabbage.

Epidemiological investigations could also be improved if the foods chosen for nutrient analysis were representative of those foods consumed by the study population. For example, the colon cancer studies identified in table 1 were conducted in numerous regions all over the world. However, because there has been no systematic sampling frame, it is difficult to determine how well the values in food composition tables represent various regional and national food supplies. Note that the nutrient content of the varieties of cabbage indigenous to various world regions may differ. Consider the differences in fibre and vitamin A content for the four different kinds of cabbage shown in table 2. There is more than a fourfold difference in the fibre content between the savoy cabbage and the Chinese and spoon varieties, while the vitamin A values ranged from 40 IU per 100 g for red cabbage to 3100 IU per 100 g for spoon cabbage. Thus, food sampling is a key issue in developing food composition tables, especially since the world supply is constantly expanding and the product on offer changing. Sampling should include new strains of edible plants and animals [10]. At present, none of the major food composition tables are based on sampling that is representative of the foods offered to consumers in defined geographic regions. Instead, data is compiled from food industry, government and independent laboratories, and from the scientific and technical literature, with each covering a different geographic area [2, 10, 25, 34, 38, 39]. The current practice is to weight the averaged analytical values of foods that are similar but not identical. Weighting schemes reflect geographic production of samples, and seasonal availability or production figures [38]. However, without a representative sample of the food at the retail level, weighted models for many foods must remain empirical. Systematic sampling is also required to determine the variance of nutrients in foods consumed by specific regional populations.

As shown in table 3, food sampling variance differs from one nutrient to another and from food to food. For example, note that the standard error for the mean is large for the iron content of apples selected from the retail food supply of Utah, but is small for the fibre content. A statistically significant nutrient difference in consumption may be observed between two groups, but these differences are not meaningful if the difference is less than the food sampling variance.

In addition to regional, seasonal, and maturational variations and differences between various parts of a foodstuff, variability in reported food composition data may also be caused by differences in analytical method. The new methodological advances in the field of nutrient analysis, including widespread use of radio-immuno-assay (RIA), radiobioassay, fluorometry, atomic absorption, neutron activation, high-performance liquid chromatography (HPLC), stable isotope electrophoresis, and auto-analysis techniques, among others, are creating masses of new data which need to be rapidly incorporated into existing data bases if these are to be kept current and relevant.

However, there are differences in reported food composition data due to intra- and inter-laboratory variance even when samples are assayed by the same analytical techniques. Such analytical errors could bias study outcomes. The difference in the fibre content of foods analysed by three different analytical methods (shown in table 4) illustrates the point. Note that crude fibre values generally underestimate dietary fibre as measured by the newer assays. Neutral detergent fibre values from two sources, Van Soest and Mahoney, show inter-laboratory variation. It should be noted that assays were performed on different food samples. However, inter-laboratory differences between different food samples are less than those observed between different analytic methods. Such differences point out the need for suitable standard reference materials that can be distributed to laboratories as part of the quality-control process.

The effects of processing may significantly alter the nutrient content of foods. Processing includes harvesting, mechanical and heat treating, packaging, and storage procedures. Processing food products together also alters the nutrient content of the products. For example, deep frying potatoes in vegetable oil increases the fat content of the product as eaten. Table 5 shows the effects of boiling on selected cruciferous vegetables. Although calories remain constant, vitamins A and C decrease with cooking.

Cruciferous vegetables have characteristics as a group that appear to be protective for some types of cancer, including bowel cancer. Note, however, that the nutrient content of these vegetables is quite different. Thus, using cruciferous vegetables as a class reduces the quantitative power of a study unless the proportions of the individual vegetable consumed are known; one should therefore document manipulations of collapsing of data in a data base used for specific studies.

Going one step further, the potential protection of crucifers may be conferred by the non-nutrient compounds, aromatic isothiocyanates. There is no quantitative information about the concentration of these compounds in foods, but there is no reason to believe that their levels in foods are any more constant than those of essential nutrients. The study of the relationship of diet to health and disease may therefore

Table 3. Selected examples: Nutrient composition of foods in Utah (amount per 100 g edible material). Retail food sample variance

Food and description	Index number	Water[a] (g)	Fat (g)	Protein (g)	Neutral detergent fibre (g)	Iron (µg)	Copper (µg)	Zinc (µg)	Mn (µg)	Ash (g)
Apples Raw, commercial varieties: not pared	13	87.9 ±0.75 (87.1–88.7)	0.72 ±0.35 (0.33–1.1)	0.21 ±0.04 (0.16–0.25)	1.1 ±0.08 (0.94–1.2)	433 ±188 (197–717)	58 ±13 (43–78)	13 ±16 (0.00–40)	117 ±50 (40–175)	0.20 ±0.03 (0.16–0.25)
Apple sauce, canned Sweetened	29	77.0(4) ±1.53 (74.9–78.6)	0.49 ±0.92 (0.00–1.87)	0.14 ±0.01 (0.13–0.17)	0.71 ±0.08 (0.62–0.81)	522 ±257 (160–761)	55 ±11 (40–60)	55 ±18 (31–74)	55 ±4.2 (50–60)	0.17 ±0.3 (0.12–0.21)
Apricots Raw	30	85.2 ± 5.46 (77.4–89.9)	0.39 ±0.38 (0.09–1.01)	1.86 ±1.50 (1.08–4.54)	1.32 ±0.33 (0.77–1.59)	1,220 ±510 (137–1,990)	164 ±115 (68–358)	245 ±262 (73–608)	175 ± 43 (135–217)	1.03 ±0.35 (0.66–138)
Apriots Canned, heavy syrup: drained solids		72.9 ± 2.46 (69.9–75.9)	0.56 ±0.43 (0.00–1.2)	0.94 ±0.19 (0.62–1.1)	0.81 ±0.20 (0.61–1.1)	678 ±351 (476–1,300)	114 ± 36 (69–158)	190 ± 46 (147–258)	90 ±49 (27–156)	0.29 ±0.17 (0.00–0.46)
Asparagus Canned spears: green, regular pack: drained solids	52	94.0(4) ±1.3 (92.7–95.2)	0.64 ±0.12 (0.49–0.80)	2.14 ±0.29 (1.85–2.63)	1.06 ±0.08 (0.99–1.43)	1,830 ±1,570 (601–3,960)	96 ±23 (68–117)	404 ± 45 (340–477)	175 ±132 (97–408)	0.76 ±0.54 (0.28–1.25)
Asparagus Frozen spears: cooked, boiled, drained	63	92.4(4) ±0.31 (91.9–92.6)	0.41 ±0.15 (0.29–0.64)	2.95 ±0.21 (2.68–3.17)	1.1 ±0.23 (0.82–1.4)	638 ±131 (546–830)	170 ± 10 (161–180)	556 ±114 (431–707)	172 ± 35 (131–214)	0.60 ±0.08 (0.53–0.73)
Bacon, cured Cooked, broiled or fried, drained	126	14.9 ± 5.33 (11.1–24.3)	44.0 ± 5.91 (34.9–49.4)	31.6 ± 6.83 (25.6–42.3)		2,180 ±640 (1,500–2,960)	360 ± 78 (245–458)	3,140 ±574 (2,430–4,010)	103 ± 18 (77–119)	7.50 ±1.64 (5.11–9.72)
Bananas, raw Common	141	77.7(6) ±3.00 (73.9–82.1)	0.70 ±0.66 (0.22–1.9)	0.96 ±0.17 (0.73–1.1)	1.1 ±0.48 (0.68–1.8)	382 ± 38 (346–452)	130 ± 20 (111–167)	155 ± 18 (131–184)	152 ± 44 (98–208)	0.74 ±0.07 (0.61–0.84)

a. Data reported as mean, standard error of the mean (SEM) and the range.

49

Table 4. Fibre content of various foods by different analytic methods (amount per 100 g)

	Crude fibre (g)[a]	Dietary fibre (g)[b]	Neutral detergent fibre (g)[c]	
			1	2
All bran cereal	7.80	26.7	32.98	–
Whole wheat bread	1.60	8.50	1.55	2.60
Apple	0.40	3.71	0.89	1.10
Broccoli	1.50	4.10	1.34	1.42
Cabbage	0.80	2.83	1.11	1.12
Potato	0.50	3.51	2.33	0.67

a. US Department of Agriculture, "Composition of Foods: Crude Fiber," *Agriculture Handbook No. 8* (Science and Education Administration, USDA, Washington, D.C., 1983).
b. D. A. T. Southgate, "Dietary Fiber," *J. Hum. Nutr.*, 30: 303 (1976).
c. Neutral detergent fibre: (1) P. J. Van Soest, "Fiber Analysis Table: By the Amylase Modification," *A.J.C.N.*, 31: s281–s284 (1978); (2) A. W. Mahoney, S. K. Collinge, B. H. Byland, and A. W. Sorenson, *Nutrient Composition of Foods Contained from Retail Outlets in Utah*, Utah Agricultural Experiment Station, Research Report, 53 (Utah State University, 1980).

Table 5. Selected nutrient composition of cruciferous vegetables[a] by different cooking methods (amount per 100 g)

Type	Raw					Cooked (boiled, drained)				
	Kcal	Fibre (g)	Fat (g)	Vit. A (IU)	Vit. C (m)	Kcal	Fibre (g)	Fat (g)	Vit. A (IU)	Vit. C (m)
Cabbage (common)	24	0.80	0.18	126	47	21	0.60	0.25	86	24
Cauliflower	24	0.85	0.18	16	72	24	0.82	0.17	14	55
Brussels sprouts	43	1.51	0.30	883	85	39	1.37	0.51	719	62
Broccoli	28	1.11	0.35	1,542	93	29	1.20	0.28	1,409	63

a. Colon cancer protection may be conferred by a non-nutrient(s) component of food, aromatic isothiocyanates. There is no quantitative information on the proposed protective agent.
Source: US Department of Agriculture, "Composition of Foods: Raw, Processed, Prepared," *Agriculture Handbook No. 8–11* (Science and Education Administration, USDA, Washington, D.C., 1984).

require accurate information on the non-nutrient as well as the nutrient components of food. However, there is very little non-nutrient food composition data available.

In addition to non-nutrient data, future epidemiologic research will demand more information on subunits of nutrients, including data on biologically active forms of compounds found in food with chemically or physically different components. Pyridoxal/pyridoxine dietary fibre and carotenoids and retinoids are examples of difficult biological forms of nutrients.

A major hindrance to epidemiologic studies are data sets with missing values. Even though a computerized system is designed to update and expand food composition data, there will always be incomplete nutrient or food information which will necessitate users' judgement for dealing with missing data. Epidemiologists, like other users, are forced to fill in "zeros" in data bases with imputed values. Estimates of missing values may come from data on similar items, recipe calculations, or even values based on educated guesses. Raw values are often substituted for food usully consumed cooked, for example in relation to meat. And sometimes a food or food group is used as a surrogate for the nutrient content of diets: milk has been used to estimate retinoid values of diets while selected fruit and vegetables have been used as estimators of beta-carotene. Decisions regarding inputting missing values would be better made on standardized criteria developed by panels of experts in the fields of nutrition and data base management.

SUMMARY

Some illustrative problems related to nutrient data bases that have the potential to affect the outcomes of epidemiological research have been presented here. This list is by no means exhaustive of all problems that are encountered in this kind of diet-related research. In general, additional food composition data will improve the power of many epidemiologic research projects while standardization and careful documentation of data bases will allow more appropriate comparisons between studies. Bioavailability, nutrient (and non-nutrient) interactions, and the influence of environmental factors on food composition all have an impact on the outcome of diet-related epidemiologic investigations. These factors represent new challenges in nutrient data base management.

REFERENCES

1. S. Bingham, R. R. Williams, J. J. Cole, and W. P. T. James, "Dietary Fibre and Regional Large-bowel Cancer Mortality in Britain," *Br. J. Cancer*, 40: 456–463 (1979).
2. R. R. Butrum and S. E. Gebhardt, "Nutrient Data Bank: Computer-based Management of Nutrient Values in Foods," *J. Am. Oil Chemist' Soc.*, 53 (1976).
3. J. F. Calder, M. W. Wachira, T. Van Sant, M. S. Malik, and R. N. Bowry, "Diverticular Disease, Carcinoma of the Colon and Diet in Urban and Rural Kenyan Africans," *Diagn. Imaging*, 49: 23–28 (1980).
4. B. M. Calkins, D. J. Whittaker, P. P. Nair, A. A. Rider, and N. Turjman, "Diet, Nutrition Intake, and Metabolism in Populations at High and Low Risk for Colon Cancer: Nutrient Intake," *A.J.C.N.*, 40: 896–905 (1984).
5. L. B. Dales, G. D. Friedman, H. K. Ury, S. Grossman, and S. R. Williams, "A Case-control Study of Relationships of Diet and Other Traits to Colorectal Cancer in American Blacks," *Am. J. Epidemiol.*, 109: 132–144 (1979).
6. H. N. Englyst, S. A. Bingham, H. S. Wiggins, et al., "Nonstarch Polysaccharide Consumption in Four Scandinavian Populations," *Nutr. Cancer*, 4: 50–59 (1982).
7. J. E. Enstrom, "Colorectal Cancer and Beer Drinking," *Br. J. Cancer*, 35: 674 (1977).
8. S. Graham, H. Payal, M. Swanson, A. Mittleman, and G. Wilkinson, "Diet in the Epidemiology of Cancer of the Colon and Rectum," *J. Natl. Cancer Inst.*, 61: 709–714 (1978).
9. W. Haenszel, H. B. Locke, and M. Segi, "A Case-control Study of Large Bowel Cancer in Japan," *J.N.C.I.*, 64: 17–22 (1980).
10. R. G. Hansen, B. W. Wyse and A. W. Sorenson, *Nutritional Quality Index for Food* (AVI, Westport, Conn., 1979).
11. M. W. Hinds, L. N. Kolonel, J. Lee, and T. Hiroshata, "Associations between Cancer Incidence and Alcohol/Cigarette Consumption among Five Ethnic Groups in Hawaii," *Br. J. Cancer*, 41: 929–940 (1980).
12. International Agency for Research on Cancer Intestinal Microecology Group, "Dietary Fiber, Transit-time, Fecal Bacteria, Steroids, and Colon Cancer in Two Scandinavia Populations," *Lancet*, 2: 207–211 (1977).

13. M. Jain, G. M. Cook, F. G. Davish, M. G. Grace, G. R. Howe, and A. B. Miller, "A Case-control Study of Diet and Colo-rectal Cancer," *Int. J. Cancer*, 26: 757–768 (1980).

14. O. M. Jensen, "Cancer Risk among Danish Male Seventh Day Adventists and Other Temperance Society Members," *J.N.C.I.*, 70: 1011–1014 (1983).

15. O. M. Jensen and R. MacLennan, "Dietary Factors and Colorectal Cancer in Scandinavia," *Isr. J. Med. Sci.*, 15: 329–334 (1979).

16. O. M. Jensen, R. MacLennan, and J. Wahrendort, "Diet, Bowel Function, Fecal Characteristics, and Large Bowel Cancer in Denmark and Finland." *Nutr. Cancer*, 4: 5–19 (1982).

17. L. J. Kinlen, "Meat and Fat Consumption and Cancer Mortality: A Study of Strict Religious Orders in Britain," *Lancet*, 1: 946–949 (1982).

18. L. K. Kolonel, J. H. Hankin, A. M. Nomura, and S. Y. Chu, "Dietary Fat Intake and Cancer Incidence among Five Ethnic Groups in Hawaii," *Cancer Res.*, 41: 3727–3728 (1981).

19. P. A. Kurup, N. Jayakumari, M. Indira, et al., "Diet, Nutrition Intake, and Metabolism in Population at High and Low Risk for Colon Cancer: Composition, Intake, and Excretion of Fiber Constituents," *A.J.C.N.*, 40: 942–946 (1984).

20. K. Liu, J. Stamler, D. Moss, D. Garside, V. Persky, and L. Soltero, "Dietary Cholesterol, Fat, and Fiber, and Colon-cancer Mortality," *Lancet*, 2: 782–785 (1979).

21. J. L. Lyon and A. W. Sorenson, "Colon Cancer in a Low-risk Population," *A.J.C.N.*, 31: s227–s230 (1978).

22. R. MacLennan, O. M. Jensen, J. Mosbech, and H. Vuori, "Diet, Transit Time, Stool Weight, and Colon Cancer in Two Scandinavian Populations, *A.J.C.N.*, 31: s239–s242 (1978).

23. S. L. Malhotra, "Dietary Factors in a Study of Colon Cancer from Cancer Registry," *Med. Hypotheses*, 3: 122–126 (1977).

24. O. Manousos, N. E. Day, D. Trichopoulos, F. Gerovassilis, A. Tzonou, and A. Polychronopoulai, "Diet and Colorectal Cancer: A Case-control Study in Greece," *Int. J. Cancer*, 32: 1–5 (1983).

25. R. A. McCance and E. M. Widdowson, eds., *The Composition of Foods*, 4th ed. (HMSO, London, 1978).

26. A. J. McMichael, M. B. McCall, J. M. Hartshorne, and T. L. Wooding, "Patterns of Gastrointestinal Cancer in European Migrants to Australia: The Role of Dietary Change," *Int. J. Cancer*, 25: 431–437 (1980).

27. A. J. McMichael, J. D. Potter, and B. S. Hetzel, "Time Trends in Colo-rectal Cancer Mortality in relation to Food and Alcohol Consumption: US, United Kingdom, Australia, and New Zealand," *Int. J. Epidemiol.*, 8: 295–303 (1979).

28. A. B. Miller, G. R. Howe, M. Jain, K. L. P. Craib, and L. Harrison, "Food Items and Food Groups as Risk Factors in a Case-control Study of Diet and Colo-rectal Cancer," *Int. J. Cancer*, 32: 155–161 (1983).

29. B. Modan, "Patterns of Gastrointestinal Neoplasms in Israel," *Is. J. Med. Sci.*, 15: 301–304 (1979).

30. L. W. Pickle, M. H. Greene, R. G. Ziegler, A. Toledo, R. Hoover, H. T. Lynch, and R. F. Fraumeni, Jr., "Colorectal Cancer in Rural Nebraska," *Cancer Res.*, 44: 363–369 (1984).

31. J. W. Powles and D. R. R. Williams, "Trends in Bowel Cancer in Selected Countries in relation to Wartime Changes in Flour Milling," *Nutr. Cancer*, 6: 40–48 (1984).

32. R. W. Rawson, "The Total Environment in the Epidemiology of Neoplastic Disease: The Obvious "Ain't Necessarily So," *Cancer Incidence in Defined Populations*, Banbury Report, no. 4 (Cold Spring Harbor Laboratory, Cold Spring Harbor, N.Y., 1980), pp. 109–119.

33. B. S. Reddy, G. Ekelund, M. Bohe, A. Engle, and L. Domellof, "Metabolic Epidemiology of Colon Cancer: Dietary Pattern and Fecal Sterol Concentrations of Three Populations," *Nutr. Cancer*, 5: 34–40 (1983).

34. R. L. Rizek, B. P. Perloff, and L. P. Posati, "USDA's Nutrient Data Bank," *Food Tech. in Australia*, 33: 3 (1981).

35. P. Rozen, S. M. Hellerstein, and C. Horwitz, "The Low Incidence of Colorectal Cancer in a High-risk Population," *Cancer*, 48: 2692–2695 (1981).

36. G. N. Stemmermann, A. M. Y. Nomura, and L. K. Heilbrum, "Dietary Fat and the Risk of Colorectal Cancer," *Cancer Res.*, 44: 4633–4637 (1984).

52

37. A. R. P. Walker, "The Relationship between Bowel Cancer and Fiber Content in the Diet," *A.J.C.N.*, 31: s248–s251 (1978).
38. B. K. Watt, "Tables of Food Composition: Uses and Limitations," *Contemp. Nutr.*, 5: 2 (1980).
39. B. K. Watt and A. L. Merrill, *Composition of Foods: Raw, Processed, Prepared* (USDA, Washington, D.C., 1963).
40. R. Zaldivar, W. H. Wetterstrand, and G. L. Ghai, "Relative Frequency of Mammary, Colonic, Rectal, and Pancreatic Cancer in a Large Autopsy Series," *Zbl. Bakt. B,* 169: 474–481 (1979).

5
Epidemiological Uses of Food Composition Data in the European Context

LENORE ARAB

Klinisches Institut für Herzinfarktforschung, Heidelberg, Federal Republic of Germany

INTRODUCTION

Instead of presenting a long list of users and potential users of food composition data in Europe, and an even longer list of the needs that are not yet fulfilled in that field, this paper focuses on one specific user group and their problems – the nutritional epidemiologists. A close look at the difficulties faced by persons conducting international nutritional epidemiology sheds light on many of the European and non-European problems in the use of food composition data.

Relevant to the broader context of users and needs of food composition data in Europe is the collection of issues presented at the workshop entitled "Towards Compatibility of Food Nutrient Composition Data Bases in Europe" [22]. At this meeting, held in Wageningen, the Netherlands, EUROFOODS was founded and work committees established to direct their efforts towards solving European food problems in five areas. Responses to a questionnaire circulated in conjunction with this meeting revealed the breadth and depth of food-nutrient data bases in Europe, and their limitations [3]. A recent update on the number of food entries in the systems of various European countries can be found in figure 1.

NUTRITIONAL EPIDEMIOLOGY

In the attempt to determine and quantify nutritional factors related to the onset of disease in the human population, epidemiologic analyses are being applied in cross-sectional, case-control, and cohort studies (see chap. 6). Relationships between the eating behaviour of individuals and the onset or prevalence of risk factors and chronic diseases are being examined and quantified. Conducting such studies internationally, by broadening the range of observation, enhances the ability to identify risk differentials and thereby strengthens the conclusions drawn from population studies.

Nutritional epidemiology incorporates the attempt to collect, systematically and

54

in an unbiased manner, information on the eating behaviour of individuals and groups for use in illuminating or testing correlates and causality of disease.

THE PROBLEMS

The difficulties of nutritional epidemiology as they relate to the nutritional side begin with problems in dietary assessment methodology [9] (see chap. 10). This includes the questions asked, how they are asked, who asks them, built-in validation attempts, the degree of specificity strived for, the forms, booklets, and guidelines used, the conditions surrounding the information exchange, and motivational measures, among other factors. In addition to these difficulties, quantification of amounts [8, 13, 21], translational difficulties, coding of consumed foods into computer-readable form [2], differences between tables [17] and between foods, and what to do about missing values present trying and complex obstacles. Some of these issues are related directly, others indirectly, to food composition tables.

These problems are more clearly seen in perspective against the background of a typical dietary assessment. Taking, for example, the often-used 24-hour recall method to compare intakes in high-risk individuals and a control group, the following steps, listed with their possible shortcomings, are necessary:

1. Question the individual about what was eaten the previous day. (Memory failure; interviewer-related bias.)
2. Question the individual about the amounts of the foods eaten. (Memory failure; estimation difficulties; non-edible portions – bones, pits, skin.)
3. Code foods into machine-readable form, usually numbers. (Few codes leading to compromised information; errors in reading or writing codes.)
4. Convert portion sizes into gram amounts. (Plate waste, refuse deduction, portion size calculation from recipes.)
5. Enter subject identification, date, meals, foods, and amounts. (Transcription errors.)
6. Check entered data for correctness. (Oversight; difficult forms; no printout of food names.)
7. Correct the errors. (Renewed typing errors.)
8. Recheck the entered corrections (Oversight or elimination of this step altogether.)
9. Merge this information into a common data base with food-nutrient information and calculate the average nutrient intake for the day. (Non-standard algorithms; program errors.)
10. Group foods for comparison of frequency and amounts consumed of basic food groups between individuals. (Double counting; mistakes in grouping.)
11. Compare intake of nutrients between case and control groups. (Missing values in nutrient tables resulting in artifactual differences.)
12. Test for significant differences between groups. (Invalid methods selected.)

Every step in this process is fraught with potential errors. Questioning about previous intake varies from interviewer to interviewer [16] unless, as is done in some cases, the subject is asked to reconstruct on paper his meals of the previous day [1] or a computer program prompts questions [6]. The subject may inadvertently adjust responses to cues from the interviewer. Subjectivity in recording responses is also

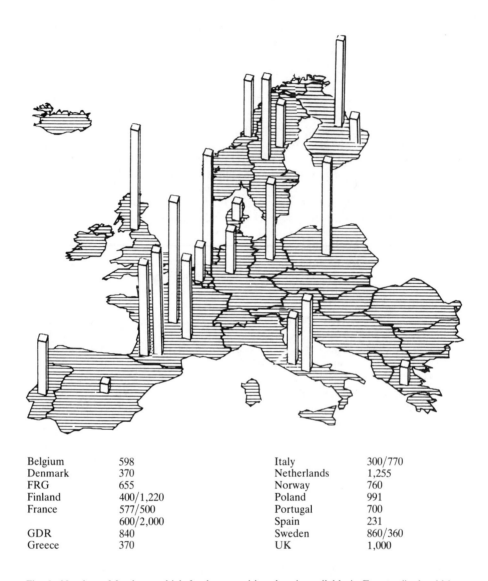

Belgium	598	Italy	300/770
Denmark	370	Netherlands	1,255
FRG	655	Norway	760
Finland	400/1,220	Poland	991
France	577/500	Portugal	700
	600/2,000	Spain	231
GDR	840	Sweden	860/360
Greece	370	UK	1,000

Fig. 1. Number of foods on which food composition data is available in Europe (it should be noted that nutrient information is often borrowed; the sum is therefore not representative of the total amount of original data).

a danger. Estimation of the amount eaten is subject to memory failure [16], estimation difficulties, and misquantification or incorrect subtraction of non-edible portions [15]. The coding of foods usually involves a compromising of the available information to fit the length and breadth of the coding system used (see chap. 12). The coding process itself, unless computerized [18], generally involves searching for the correct number from code books, transcribing it onto paper, and having these numbers entered into a terminal. This process allows for many possible reading, transcribing, and keyboarding errors. Different individuals may code the same information differently [9].

The calculation of food intakes into nutrients depends on the availability of information on the foods as consumed (cooked, fried, etc.) for the nutrients of interest. Missing values in food tables are often handled interchangeably as zero values for these nutrients, which can result in false or artifactual results [4]. Inaccuracies in the calculation of nutrients from foods can also result from faulty programming, although this is seldom a major difficulty. Differences in estimating nutrient intakes have been documented between systems with identical sources of nutrient information, for a number of reasons [11].

Regrouping the tens of thousands of food items on the market in a particular country is also generally non-standardized. Comparisons are therefore subject to differences in the systems used [16]. An example of this is butter being included in either the dairy group or the fats and oils group, and egg-rich products such as quiches and egg noodles being grouped together with cereal and grain products.

Mention should also be made of the misuse of methods for the testing of hypotheses as a major problem in nutritional epidemiology. For example, the results of a single 24-hour recall or a frequency questionnaire of food consumption are often used to determine interrelationships and the interpretation made that no underlying relationship exists. Strong interactions may be statistically insignificant due to low subject number or days of observation and great intra-individual day-to-day variability [5, 7].

SUGGESTIONS FOR IMPROVEMENT

Since the use of food composition tables in nutritional epidemiology concerns mainly estimation of the composition of food consumed by individuals, it makes sense to regard the entire process as closely interrelated – the dietary assessment, food coding, and referencing of tables. Some of the problems beginning with the assessment of dietary behaviour are: (1) non-standardized methodology; (2) portion size estimation; (3) translation; (4) mixed dishes; (5) different food codes; (6) different food grouping systems; (7) differences in food tables; (8) food compositional differences; (9) missing values; and (10) definitions of edible parts.

In many cases only minimal effort would be required to improve and simplify nutritional analysis in international epidemiological studies. The components of food-nutrient data-base systems (data files and programs) are compartmentalized and presented diagramatically in figure 2, emphasizing the individual areas needing enhanced development as well as their interdependence. The numbers in the diagram link relevant system components to numbered suggestions in the following text.

1. Standardized Dietary Assessment Methods

Although basically common terminology is used in dietary assessment, the procedures followed are non-standardized and seldom described or documented. A 24-hour recall for one person may be a computer-based decision tree [6], for another it might be an empty piece of paper, a pencil, and a simple question on the subject "Please write down what you ate yesterday." Handbooks on standardized procedures for different dietary assessment methods would improve the situation.

2. Standardized Portion Sizes

All subjects have difficulties in estimating the exact portion sizes of foods consumed (particularly retrospectively), requiring estimations and assumptions to be made by the interviewer or coder. The difference between weight of purchased product (often known) and that of actual item consumed (without bone, pits, skin, with water, etc.)

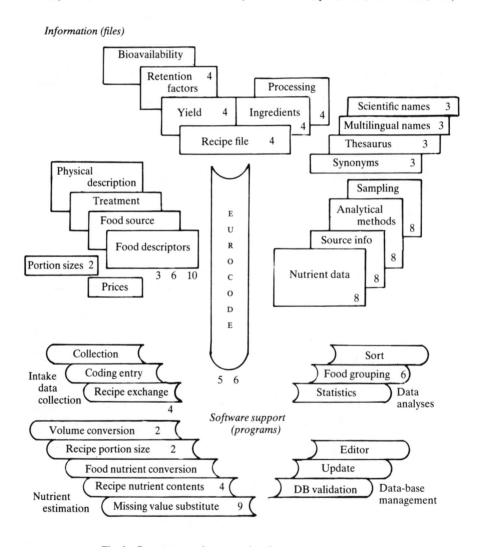

Fig. 2. Components of an extensive dietary assessment system.

Table 1. Native-language food-table entries for apples, butter, beer, and beef

English	Apple	Butter	Beer	Beef
Danish	Æble	Smør	Øl	Oksekød, højrebsfilet, afpudset, rå
Spanish	Manzana	Mantequilla	Cerveza	Carne, semigrasa (vacuno)
German	Apfel	Butter	Bier	Rindfleisch, Fehlrippe
Dutch	Appel	Boter	Bier	Rundvless, gemiddeld, vet onbereid
Italian	Mela	Burro	Birra	Bovino adulto semigrassa
Greek	Mila	Boutyro	Bira	Bodino, apaxes mbriz.omes
Swedish	Äpple	Smör	Maltdryck	Nöt biff dubbel
Portuguese	Maçã	Manteiga	Cerveja	Vaca
Finnish	Omena	Voi	Keskiolut	Naudanliha paisti

are all part of this issue. The use of standardized portion sizes for individual foods within countries could help. The application of standard weights and density measures where applicable, as well as common models for small, medium, and large portions, cuts of meat, portions of grains, noodles, and baked products, would be useful. A standard set of food models for use in dietary assessment could be a step in the right direction.

3. Names and Translations

The difficulty of translating foods from one language to another cannot be underestimated. Even within a country, colloquial food names are inexact and can be misleading. Table 1 gives a comparison of the differences between countries in the names of simple foods in food tables. Here apples, butter, beer, and beef, as entered in the food tables of 10 countries, are compared. Apples and butter between countries are somewhat comparable. Beer, however, brings in the complication of different products with different formulations and alcohol content. And beef is even more troublesome, as different countries have different standards for cutting and names for the various cuts, different fat contents and extent of trimming. A considerable part of an international coding workshop held in Heidelberg in February 1985 was devoted to the description of food items. (Unpublished proceedings available from the author.) Table 2 presents a list of many of the published food nutrient tables currently available in Europe. To use these tables one needs not only a good translation of the food items and measured parameters, but also an understanding of the language to read the introduction in which critical information on using the tables is found. A common pitfall in translating foods is in situations where an identical or similar food name is used in two countries but an entirely different product is meant. Another problem is that of foods whose names convey little meaning as to content, such as those found in table 3.

4. Mixed Dishes

People often eat dishes that are prepared elsewhere and the identity of whose ingredients remains unknown to them. Salads, stews, soups, cakes, casseroles – the list of foods in which one cannot determine what has been added just by looking is practically endless. Mixed dishes in general present problems, as hundreds of different recipe variations are involved for each dish of a given name. They are such an important

Table 2. Major European food tables

Belgium	1984	Lambin
Denmark	1983	Møller
Federal Republic of Germany	1982/83	Souci, Fachmann, Kraut
Finland	1980	Koivoistoinen
	1981	Varo
	1983	Turpeinen
France	1979	Renaud
	1980	Ostrowski and Josse
	1980	Randoin
German Democratic Republic	1979	Haenel
Greece	1982	Trichopoulou
Italy	1981	Fidanza and Versiglioni
	1983	Carnovale and Muccio
Netherlands	1984/85	UCV Tables
Norway	1977	Landsforeningen for Kosthold og Helse
Poland	1978	Berger et al.
Portugal	1977	Gonzales-Ferreira, da Silva Graza
Spain	1983	Andujar Anas, Moreiras-Verela, Extremara
Sweden	1981/84	Statens Livsmedelswerk
United Kingdom	1978/85	Paul and Southgate

Table 3. Some mixed dishes translated into English

Netherlands (Dutch)	Children's happy biscuits
	Filet américain
Denmark (Danish)	Italian salad
United Kingdom (English)	Welsh rarebit
Federal Republic of Germany (German)	Lady's finger
	Bread "Steinmetz"
Finland (Finnish)	Jansson's temptation
Italy (Italian)	Cake Margherita
Poland (Polish)	Bird milk with lemon
	Cod Greek-style
	Popular hunter stew
Sweden (Swedish)	Pyramid cake
USA (American)	Hamburger

part of European diets, however, that they cannot be ignored. A simple example is pizza, which is known everywhere by this name but is prepared in a myriad of ways – in some areas it is made without tomatoes, in others without cheese, and when cheese is used sometimes it is mozarella, sometimes Gouda, sometimes emmentaler. Standard recipes would at least prevent arbitrary differences and allow for calculations on a common basis when exact information is not available.

Despite the fact that the large proportion of the European diet comes from mixed dishes, relatively few prepared dishes have been analysed and presented in composition tables [3]. There are different methods for imputing the nutrient values of food that have undergone various preservation and preparation processes, none of which has been shown to be the most accurate [14]. Recommended procedures for imputing values would at least enhance the comparability of results.

5. A Standard Code

Different coding systems with different orientations, different groupings of food, and different levels of specificity [2] stand in the way of data exchange, comparison of results, and recalculation of surveys based on foreign tables. A common coding system would allow comparisons between studies of the frequencies and amounts of individual foods consumed. Also, it would help bypass the translational problems in identification of the same and similar foods. It would also simplify the application of common grouping schemes for foods.

6. Basic Food Groups

The problem of grouping foods and products has been addressed earlier. Since tens of thousands of food items are consumed in Western populations, it is necessary to aggregate individual foods into groups for further analyses. Brand-name items usually fall together, but egg dishes, vegetable casseroles, and stews are foods that typically differ from one grouping scheme to another. Different schemes for grouping foods can lead to artifactual differences or false conclusions regarding epidemiologic associations. Agreement on at least the major food groups and guidelines for alternate grouping schemes would be useful.

7. Food-table Format

Differences between different food tables are apparent as soon as one opens two different tables. The format with which foods and nutrients are listed differs, some tables have one food per page, others lists of foods and lists of nutrients. The ordering of the nutrients differs, as do the units of measure, the exactness, and the number of significant digits. There are also differences in the factors used for calculation [4]. Again, the need for a standardized format is demonstrated here (see chap. 10).

8. Food-nutrient Differences

The nutrient composition of an apple or potato in the FRG is not likely to be the same as in Holland, and is even less likely to be the same as in Sweden or in Spain. First of all, different species are used; second, different amounts of sunshine influence the nutrient level; third, the trace elements are strongly influenced by the soil. The breadth of variation may be as great within a country as between countries. The fat content of similar foods differs tremendously from country to country. Little can be done here in general, but this problem discourages the exchange of data, particularly important when large amounts of the products are being exchanged within the European Common Market or eaten on vacation outside the country. Data exchange requires some means for the user to find the proper food in a language he can understand (source and sampling information) and a presentation of the values whose analytical method and accuracy are available to him. Since variability of values within a table can be considerable, information is needed to enable the end-user to know the laboratory intra-sample variation and variation between species, as well as the origin of the food sampled and the time of analysis.

9. Missing Values

There are many types of missing values in nutrient tables. As Greenfield and Southgate have pointed out, there are imputed values, borrowed values, values calculated from a measured factor, values that are available for many foods but not available for others (absent), and nutrients that are not included at all [19]. For the epidemiologist the values which are labelled missing bring further analyses to a halt. Zeros in the analyses make the determination of a gradient of risk impossible – an estimation is almost always preferable. Greater exchange of nutrient information between countries would allow for the eradication of many of the missing values in tables. Efforts to correlate information on nutrient losses and gains in preparation of foods would help improve the quality of estimation of nutrient contents in preferred foods which have not been analysed.

10. Edible Parts

So-called "edible parts" of foods can differ from culture to culture. The French for example consume only the soft inner portion of Camembert cheese. Germans eat the crust as well, and would not expect this to be missing in the nutrient analyses. Descriptions of the food sample analysed would help prevent misuse of foreign data.

This list of possible problems with food composition data is not exhaustive. Additional suggestions for improvement would include increased availability of data [3] in machine-readable form and a solution for updating data more easily and less expensively.

ONGOING ACTIVITIES

To address a number of the problems mentioned here a project has been awarded to EUROFOODS by the European Community. This involves co-operative efforts between 17 countries towards the development of a merged European food composition data bank. The five aspects of the pilot project are:
1. Inventorization of existing food-nutrient data bases, and examination of the needs of current and potential users of European nutrient data bases.
2. Examination of the differences between national food composition tables, and proposal of ways in which they might be better harmonized.
3. Development of a common coding system for foods for use in the harmonization of nutrient tables and collection of food-intake information.
4. Estimation of the operational requirements of a merged European nutrient data-bank system.
5. Establishment of a mechanism for consultation with member states.

Three workshops have taken place with the central theme involving the development of a common food coding and identification system. In the Federal Republic of Germany, a commission with members from four European countries which developed a German Food Code (the BLS [19]) has been actively involved since the beginning of 1984 in developing a draft for this common EUROCODE. Additionally, English translations of the foods presented in 16 of the European food composition tables are currently available in a data file in Heidelberg, along with their EUROCODE 2 number. Efforts are now being undertaken to collect and translate into English typical national dishes and their recipes.

An example of governmental encouragement of standardization in the dietary assessment methodology is the current situation in the Federal Republic of Germany. One of the major research funding agencies, the Ministry of Research and Technology, is requesting that all new projects incorporating dietary assessment incorporate as far as possible the existing standards and the BLS, the Federal Food Code.

SUMMARY

Discussions of food tables should not be divorced from dietary intake assessment methods. The users of food-nutrient information, particularly in the field of nutritional epidemiology, need a complete system with which information can be accurately and in standardized form collected, coded, and translated into nutrient composition data. The current state of affairs is such that most individuals working in this field must create their own system. The wide range of previously mentioned possibilities for variation and error practically excludes the possibility of comparing results between studies. International epidemiology suffers from all these problems, in addition to the problems of language, mixed dishes, and different foods. These users, who can potentially contribute greatly through their findings to the public health of nations through elucidation and quantification of the role of dietary behaviour on the etiology of disease, should not be neglected in the general consideration of the users of and need for food composition data.

REFERENCES

1. K. J. Acheson, I. T. Campbell, O. G. Edholm, D. S. Miller, and M. J. Stock, "The Measurement of Food and Energy Intake in Man: An Evaluation of Some Techniques," *A.J.C.N.*, 33: 1147–1154 (1980).
2. L. Arab, "Coding and Entry of Food Intakes," in R. Tobelmann, ed., *Proceedings of Eighth National Nutrient Data Bank Conference, July 25–27, 1983, Minneapolis, Minnesota* (National Technical Information Service, Springfield, Va., 1983), pp. 13–22.
3. L. Arab, "Summary of Survey of Food Composition Tables and Nutrient Data Banks in Europe," in C. E. West, ed., "EUROFOODS: Towards Compatibility of Nutrient Data Banks in Europe," *Annals of Nutrition and Metabolism*, 29 (Suppl. 1): 39–45 (1985).
4. L. Arab and M. Wittler, *A Comparision of Food Composition Tables in Europe* (in press).
5. G. H. Beaton, J. Milner, A. P. Corey, et al., "Sources of Variance in 24-hour Dietary Recall Data: Implications for Nutrition Study Design and Interpretation," *A.J.C.N.*, 32: 2546–2559 (1979).
6. O. Bellin and L. Arab, "ESSEKAN – ein Standardisiertes 24-Stunden-Erinnerungsprotokoll über körperliche Aktivitat und Nahrungsmittelaufnahme," in L. Arab and G. Karg, eds., *Entwicklung und Benutzung von Nahrstoff-Datenbanken in der Bundesrepublick Deutschland* (Bundesforschungsanstalt für Ernährung, Karlsruhe, 1984), pp. 192–203.

7. S. Bingham, "Premise and Methods," in G. G. de Backer, H. T. Pedoe, and P. Ducimetiere, eds., *Surveillance of Dietary Habits of the Population with regard to Cardiovascular Diseases*, EURO-NUT Report, no. 2 (Stichting Nederlands Instituut voor de Voeding, Wageningen, 1983), pp. 21–42.
8. A. Ferro-Luzzi, "Standardization within and between Centers," in G. G. de Backer, H. T. Pedoe, and P. Ducimetiere, eds., *Surveillance of Dietary Habits of the Population with regard to Cardiovascular Diseases*, EURO-NUT Report, no. 2 (Stichting Nederlands Instituut voor de Voeding, Wageningen, 1983), pp. 51–60.
9. G. C. Frank, A. T. Hollatz, L. S. Webber, and G. S. Berenson, "Effect of Interviewer Recording Practices on Nutrient Intake – Bogalusa Heart Study," *J. Am. Diet. Assoc.*, 84(12): 1432–1439 (1984).
10. J. G. A. J. Hautvast and W. Klaver, eds., *The Diet Factor in Epidemiological Research*, EURO-NUT Report, no. 1 (Ponsen and Looyen, Wageningen, 1982).
11. L. Hoover, "Computerized Nutrient Data Bases: I. Comparison of Nutrient Analysis Systems," *J. Am. Diet. Assoc.*, 82(5): 501–505 (1983).
12. A. Keys, "Dietary Survey Methods," in R. I. Levy, B. M. Rifking, B. H. Dennis, and N. D. Ernst, eds., *Nutrition, Lipids, and Coronary Heart Disease, A Global View* (Raven Press, New York, 1979), pp. 1–23.
13. J. P. Madden, S. J. Goodman, and H. A. Guthrie, "Validity of the 24-hour Recall," *J. Am. Diet Assoc.*, 68: 143–147 (1976).
14. A. Marsh, "Problems Associated with Recipe Analysis," in R. Tobelmann, ed., *Proceedings of Eighth National Nutrient Data Bank Conference, July 25–27, 1983, Minneapolis, Minnesota* (National Technical Information Service, Springfield, Va., 1983), pp. 29–38.
15. R. H. Matthews and Y. J. Garrison, *Food Yields Summarized by Different Stages of Preparation*, Agriculture Handbook 102, (USDA, Washington, D.C., 1975).
16. National Research Council, "Methodology," *Diet, Nutrition, and Cancer* (National Academy Press, Washington, D.C., 1982), pp. 30–50.
17. J. Perisse, "The Heterogeneity of Food Composition Tables," in J. G. A. T. Hautvast and W. Klaver, eds., *The Diet Factor in Epidemiological Research*, EURO-NUT Report, no. 1 (Ponsen and Looyen, Wageningen, 1982), pp. 100–105.
18. H. Pfannendorfer and L. Arab, "KALI-Kodierungs- und Auswertungs-programm für Lebensmittel-Inhaltsstoffe," in L. Arab and G. Karg, eds., *Entwicklung und Benutzung von Nahrstoff-Datenbanken in der Bundesrepublick Deutschland* (Bundesforschungsanstalt für Ernährung, Karlsruhe, 1984), pp. 137–147.
19. H. Rottka, "Bundeslebensmittelschlüssel (BLS)," in L. Arab and G. Karg, eds., *Entwicklung und Benutzung von Nahrstoff-Datenbanken in der Bundesrepublik Deutschland* (Bundesforschungsanstalt für Ernährung, Karlsruhe, 1984), pp. 126–129.
20. D. A. T. Southgate and H. Greenfield, *Guidelines to the Production, Management and Use of Food Composition Data Systems* (in press).
21. W. A. van Staveren and J. Burema, "Validity and Reproducibility of Methods," in G. G. de Backer, H. T. Pedoe and P. Ducimetiere, eds., *Surveillance of Dietary Habits of the Population with regard to Cardiovascular Diseases*, EURO-NUT Report, no. 2 (Stichting Nederlands Instituut voor de Voeding, Wageningen, 1983), pp. 43–50.
22. C. E. West, ed., "EUROFOODS: Towards Compatibility of Nutrient Data Banks in Europe," *Annals of Nutrition and Metabolism*, 29 (Suppl. 1): 1–72 (1985).

6
NCI Food Data Needs: Impact on Coding Systems

RITVA R. BUTRUM

National Cancer Institute, Washington, D.C., USA

INTRODUCTION

The National Cancer Institute (NCI) supports many different types of research on the relationship between nutrition and cancer, including studies in prevention, epidemiology, etiology, basic cellular mechanisms, treatment, and information dissemination. As a result of conducting this research, major deficiencies in existing food data bases have become apparent. Problems range from the unavailability of food data to questionable data validity, reliability, accessibility, and interpretation.

Food data problems are especially evident in epidemiologic studies, which focus on populations' dietary differences and their relationships to cancer incidence. International variations in diet can be distinguished by broad nutrient-food categories, but detailed information is lacking on the nutrient content of many foodstuffs consumed by different peoples and their subgroups. In addition, comprehensive data are unavailable for the non-nutrient chemical components of food, which represent protection for, or risks to, good health.

Thus, creating a comprehensive food composition data network as proposed by INFOODS, covering the spectrum of biologic activity from nutrients to contaminants, would greatly aid cancer epidemiologists, researchers, clinicians, educators, and public policy-makers. The purpose of this paper is to highlight selected NCI-supported projects which would benefit from the development of such a system. Table 1 presents users of food composition data at the international, national, local, and individual levels. The objectives of each of these researchers' projects is then discussed with their needs relative to the development of an internationally acceptable coding system.

INTERNATIONAL RESEARCH

At the international level, NCI is supporting a variety of diet, nutrition, and cancer studies involving food composition data. The first study, now in progress in Puerto Rico, is examining the exposures of staple foods to high levels of toxic substances

Table 1. NCI-supported research projects

Principal investigator	Project title	Location	Objective	Rationale
Angel A. Roman-Franco	Cancer Control Research Grant for Puerto Rico[a]	San Juan, Puerto Rico	Obtain information relating to the degree of exposure of staple foods to high levels of toxic substances, e.g. pesticides and chemical fertilizers	The bulk of calories comes from a single staple, which may be the principal vehicle for exposing the population to potential carcinogens
T. Colin Cambell	Dietary Selenium and Cancer[a]	People's Republic of China	Evaluate the role of dietary selenium as a risk factor for various cancers	*In vitro* and *in vivo* studies have demonstrated that Se supplements reduce the incidence of carcinogen-induced and spontaneous neoplasms
Li Ping	Nutrition Intervention Trial[a]	Linxian, China	Conduct intervention trials using micro-nutrients administered to oesophageal dysplasia patients and to a general population at high risk	Segments of the population are known to have poor intakes of multiple micro-nutrients, which may correlate with cancer risk
Ruth M. Hicks	Carcinogenesis in Human and Rat Bladder Tissues[a]	London, UK	Compare the effects of promoters (e.g. saccharin and tryptophan) and anti-promoters (e.g. retinoids) on previously initiated human and rat bladder organ cultures	This study will determine if the rat is an appropriate model for screening human bladder carcinogens
Olli P. Heinonen	US–Finland Studies of Nutrition and Cancer[a]	Helsinki, Finland	Determine whether beta-carotene or vitamin E supplementation is effective in preventing lung cancer in heavy smokers	Dietary beta-carotene has been inversely correlated with lung cancer incidence in epidemiologic studies
Khursheed N. Jeejeebhoy	Nutritional Assessment of Cancer Patients[a]	Toronto, Canada	Assess the efficacy of three levels of nutritional support in maintaining lean body mass of patients with lung or colorectal cancer	This study will provide insight into the nutritional aspects of patients with advancing malignancy
Abraham M. Nomura	Cancer Epidemiology of the Migrant Japanese in Hawaii[b]	Honolulu, Hawaii	Distinguish the relationships between dietary habits and incidence of cancers in a migrant population	Culture and environmental conditions have been found to influence cancer risk

Investigator	Project	Objective	Location	Comment
Ritva Butrum	Low Fat Diet in Women with Stage II Breast Cancer[b]	Identify the effect of a low-fat diet on disease relapse and overall survival in stage II breast-cancer patients	Boston, Mass.; Atlanta, Ga.; Los Angeles, Calif.; Iowa City, Iowa; Chicago, Ill.; Houston, Tex.; New York, N.Y.; Pittsburgh, Pa.	Epidemiologic and case-control studies have linked a high-fat diet to increased incidence and mortality from breast cancer
	Low Fat Diet in Women at Increased Risk for Breast Cancer[b]	Evaluate whether manipulation of dietary fat will slow or halt the "promotion" phase of carcinogenesis	Houston, Tex.; Seattle, Wash.; Cincinnati, Ohio	Reduction of dietary fat, even if done late in the promotion phase, has resulted in a reduction of breast-cancer incidence in animal models
P. J. Van Soest, J. A. Marlett	Methodology and Analysis of Fiber Components in Food[b]	Develop methods for fibre analysis and analyse total fibre and individual dietary fibre components in US foods	Ithaca, N.Y.; Madison, Wis.	A reliable data base of total dietary fibre and fibre fractions is needed for NCI studies
Phyllis E. Brown	Adenomatous Colonic Polyps, A Vitamers and MFO Induction[b]	Assess the efficacy of using beta-carotene to inhibit colonic polyps and/or colonic cancer	Chicago, Ill.	Epidemiologic and laboratory research have demonstrated the preventive effects of carotenoids
A. R. Patel	Dietary Markers for Epidemiologic Studies of Cancer[b]	Identify markers of past or present dietary exposure	To be determined	Markers will be useful in validating or conducting nutritionally focused studies in cancer epidemiology
C. Butterworth, R. Rivlin	Clinical Nutrition Research Units (CNRUs)[c]	Stimulate multi-disciplinary research, education, and training in nutrition	Birmingham, Ala.; New York, N.Y.	Co-ordinated effort, intellectual stimulation, and use of shared resources can further our knowledge of nutrition and cancer
Paul Van Nevel	Cancer Prevention Awareness Program: Nutrition and Cancer[d]	Stimulate public awareness of the relationship between specific dietary patterns and cancer risk	Bethesda, Md.	Epidemiologic and laboratory evidence provides support for interim guidelines on diet and cancer

a. Research at the international level. b. Research at the national level. c. Research at the local level. d. Research at the individual level.

such as pesticides and chemical fertilizers. Many Puerto Ricans derive a major source of their caloric intake from maize, which may be the principal vehicle for exposing the population to potential carcinogens.

NCI also is supporting a number of studies in China, where the unique distribution of nutrients in the soil provides a natural laboratory for studying disease. For example, areas of the terrain differ widely in selenium content, and evidence exists that selenium deficiency may be associated with increased cancer risk. Another prevention intervention study in China involves administering multiple vitamin/mineral preparations containing beta-carotene to oesophageal dysplasia patients. Although a rare tumour in the United States, 20 per cent of the deaths in Linxian, China, are due to oesophageal cancer. A prominent part of the population's diet consists of pickled foods, which are suspected to produce carcinogens over time. It has been suggested that the micro-nutrient deficiencies, in combination with the carcinogenic substances, lead to cancer. A second, related trial involves administering a series of micro-nutrients to a general Chinese population using a multi-factorial design.

Among the international laboratory studies being supported by NCI is a project that began by looking at the possible cancer-promoting effects of saccharin and trypto-phan in human and rat bladder tissues. The London researchers now are comparing the effects of these cancer promoters to anti-promoters such as retinoids, and are gearing up for human prevention studies.

Finland provides a unique resource for studying the intervention of beta-carotene and vitamin E in a population of heavy smokers. Because of the country's past problems with tuberculosis, its public health service requires every male smoker to have an annual chest X-ray. This practice provides an opportunity to superimpose a low-cost nutritional intervention study into a system which already provides population surveillance, and good record-keeping. The five-year, double-blind randomized trial will also include a dietary survey of persons from the original cohort who subsequently developed cancer.

With respect to cancer treatment, researchers in Toronto, Canada, are studying perturbations of various nutritional parameters in breast and lung cancer patients, and are examining whether these deficiencies are correctable with existing methods of nutritional intervention. Investigators are applying expertise in nutritional assessment and in measurement of total body nitrogen with neutron activation.

UNITED STATES STUDIES

Within the United States, NCI's nutritional focus comprises two areas: (a) diet and cancer prevention, and (b) chemoprevention. Current and proposed projects that could benefit from the development of INFOODS include a large epidemiologic investigation, which is examining dietary patterns of the migrant Japanese in Hawaii and cancer incidence. Using epidemiologic data and sera samples collected from 1965 to 1968, incident cancer cases are being evaluated. The study also will examine the relationship between future cancer incidence and intakes of dietary fat, fibre, alcohol, dietary nitrates, salt, vegetables, and other food constituents.

A multi-institution clinical trial, now under way in eight US cities, is studying the effect of a low-fat diet (i.e. 20 per cent of total calories) on women at high risk for breast cancer. This protocol will test whether manipulation of dietary fat will slow or halt the subsequent promotion phase of cancer in women who have a predisposition for breast cancer.

The need for a comprehensive food data system such as that proposed by INFOODS is shown in studies on dietary fibre at Cornell University and the University of Wisconsin. Among the problems with existing fibre data are lack of standardized analysis techniques on representative foods, incomplete data, and lack of information about the meaning of current fibre values. Van Soest and Marlett are working to develop methods for fibre analysis and to analyse total fibre and individual dietary fibre components in US foods.

Similar dietary data problems exist for the food composition of carotenoids and retinoids. NCI supports several case-control studies that seek to demonstrate that levels of vitamin A consumption affect the development of colonic adenomatous polyps and adenocarcinoma. One three-year clinical trial using beta-carotene will attempt to slow the rate of polyp development and redevelopment. Vitamin A status will be assessed through blood plasma levels and dark adaptation tests. The advantage of using colonic adenomatous polyp formation, a pre-cancerous condition, as an end-point in this trial is that the condition has a short development and redevelopment time in many patients.

In other related studies, vitamin A is being administered to persons with asbestosis, for treatment of lung mesothelioma, and to chronic and former smokers for prevention of lung cancer. Women at high risk for cervical dysplasia are taking retinyl acetate in two other trials.

A new area of NCI-sponsored research involves identifying, characterizing, and validating the dietary markers of past or present dietary exposure to cancer. Various problems exist in interpreting data derived from current nutritional assessment techniques. Moreover, the latency period of carcinogenesis poses several problems in gathering information about past dietary habits, namely that: (a) current dietary intakes may not be representative of past intakes, since dietary patterns change over time, and (b) our knowledge of the changing composition of foods is limited. Markers of both present and past dietary experience, therefore, will be especially useful in nutritionally focused studies in cancer epidemiology.

LOCAL RESEARCH

Clinical Nutrition Research Units (CNRUs) are an example of NCI-supported activities at the local level that would benefit from an integrated food data system. CNRUs are designed to create or strengthen nutrition research, training, and education through co-ordinated effort, intellectual stimulation, and use of shared resources. Each CNRU consists of: research with human subjects and populations; laboratory investigations; research training; educational programmes for medical students, house staff, practising physicians, and paramedical personnel; nutritional support services; public information activities; and shared facilities and research services.

The NCI supports two CNRUs: one at the University of Alabama in Birmingham, and the other at the Memorial Sloan-Kettering Cancer Center, New York Hospital–Cornell Medical Center, and Rockefeller University in New York. Primary areas of research at the University of Alabama concern malnutrition of hospital patients, nutrition and cancer feeding problems, vitamin status assessment, and folic acid biochemistry and metabolism. In one study, researchers found that the intakes of women with cervical dysplasia were often low in ascorbate and deficient in folate. Moreover, folate supplementation resulted in reversal of the dysplasia. Investigations at the Memorial Sloan-Kettering CNRU focus on nutrition and its relationship to cancer, immunology, burns, pharmacology, and the brain as well as metabolism, diabetes, and lipids. To help find optimal ways to prevent and treat weight loss, cachexia, and other metabolism derangements that occur with cancer, CNRU researchers are examining whether riboflavin deficiency in experimental animals affects suppression of T-cells by syngeneic tumours.

INDIVIDUAL LEVEL

At the individual level, NCI is involved with cancer information dissemination programmes, which aim to influence the general public's dietary patterns. The NCI's goal is to reduce cancer mortality by 50 per cent by the year 2000. As part of its Cancer Prevention Awareness Program, the Office of Cancer Communication (OCC) is developing a national campaign to emphasize dietary changes which reduce an individual's risk of cancer. The OCC has designed publications and public service announcements promoting increased consumption of dietary fibre and decreased consumption of fat.

USES OF FOOD COMPOSITION DATA

As evident form the above discussion, NCI-supported researchers at the international, national, local, and individual levels vary in the sophistication of food composition data they would require. Scientists involved in international projects need a system with familiar names of foods, standard units of measurement, and foreign-language translations of data to understand better the relationship between nutrient intakes and cancer incidence patterns throughout the world. If associations between certain nutrient deficients (e.g. selenium) and cancer are confirmed, INFOODS could assist international agencies in identifying foods that must be imported to satisfy nutritional requirements and help prevent cancer.

National NCI-supported studies have a need for more standardized analysis techniques to permit intra-study comparisons, particularly with regard to fibre, carotenoid, and retinoid composition. Researchers would benefit from the standardization of sampling and analysis procedures, modes of data expression and conversion factors, and other guidelines developed by INFOODS. INFOODS vocabulary that would be particularly important include origin, part, process, and stage of maturity.

At the local level, access to a food data system would not only assist CNRU laboratory and clinical investigations, it would be useful in nutrition research training, nutritional support services, and the nutrition education activities of professionals, patients, and the general public. For example, medical and surgical residents or other professionals pursuing advanced nutrition training could be instructed about the INFOODS system and its standards and guidelines relating to data gathering, storage, interchange, and usage. The information could then be utilized in their research, which would ultimately contribute to the expansion of the INFOODS network.

At the individual level, consumers must be aware of the dietary modifications they can make to reduce their risk of cancer, the second leading cause of US mortality. The valid and reliable data managed by INFOODS ultimately could be used to guide the daily food choices of the American public.

IMPLICATIONS FOR INFOODS

What are the implications of this NCI-supported research for the development of INFOODS? First, to meet the needs of NCI-supported researchers, an internationally acceptable identification and classification scheme must be developed. The goal should be a non-hierarchical vocabulary, which is flexible and allows the future addition of other factors such as treatment or processes, level of maturity, and grade or quality descriptions. Second, any system must permit the user to describe exactly the type of printout or analysis needed. The NCI-support research projects illustrate the diversity of users of food composition data and the unique requirements of each study. While some researchers need complete tables with specialized information, others are focusing on the analysis of a single nutrient.

Finally, it is recommended that an attempt be made to anticipate the features that sophisticated statistical techniques will require. Statistical interpretation of data is, of necessity, an important function of any data system for all researchers, regardless of whether their focus is cancer, cardiovascular disease, diabetes, or another disease.

SUMMARY

In summary, NCI supports various research projects, at the international, national, local, and individual levels, which depend on food composition data. Although the five levels of users vary in the sophistication of the information they require, all need access to a system of food data which is relatively simple to use and contains reliable data. Any system developed also must be sufficiently flexible to permit expansion or other modifications to accommodate new trends and needs. The internal structure of the data system should accommodate various statistical analyses and printout formats for researchers' unique study requirements. Such a system would promote international participation and co-operation and, in all probability, significantly improve research.

7

Food Composition – A Key to Dietary Appraisal and Improvement in the United States

BETTY B. PETERKIN

Human Nutrition Information Service, US Department of Agriculture, Washington, D.C., USA

INTRODUCTION

The US Department of Agriculture (USDA) has traditionally been the primary provider of food composition data in the United States. Food composition data are vital to the major responsibilities of the Department: (a) to ensure that an adequate, wholesome, and nutritious supply of food is available and (b) to provide information to help Americans select good diets [21]. USDA food composition data are also used extensively by the Department of Health and Human Services (DHHS) – the agency with responsibility for studies of diet and disease and for the nutrition-labelling programme.

USDA contributes to food and agricultural progress throughout the world. This activity and its role as the provider of US food composition data explain USDA's support of INFOODS and earlier efforts to help make food composition data available in various regions of the world.

A wide range of activities helps to fulfil USDA's mission as it relates to nutrition. Research, information, education, regulatory, and food assistance programmes are examples [6]. Each requires food composition data. This paper describes some of these activities and their food composition data needs, and outlines the means by which USDA's Human Nutrition Information Service (HNIS) attempts to meet these needs (see chap. 11).

FOOD COMPOSITION DATA NEEDS

Human Nutrition Requirements Research

USDA's Agricultural Research Service (ARS), in its five Human Nutrition Research Centers, helps to define the amounts of nutrients and other dietary substances that promote optimum growth, development, and nutritional health. ARS scientists also study factors affecting the absorption of nutrients in the body. These studies, of course, require a precise understanding of the nutrient content of the diet. Many studies analyse diets for nutrient content; others calculate the nutrient content of diets using food composition tables. The individual centres focus on different research areas. For example, the centre at Tufts University in Boston studies the nutrition factors affecting the aging process, and the one in Houston at the Baylor College of Medicine studies the nutrition of infants and children. One of the research areas of the Human Nutrition Research Center at Beltsville, Maryland, is the development of methods for determining the composition of foods (see chap. 18).

Additional research on human nutrition requirements and the nutrient content of foods and diets is conducted at 58 state agricultural experiment stations. The Co-operative State Research Service serves as a link with the state system. Major research on food components as they relate to various diseases is conducted in another federal agency, the National Institute of Health (NIH), and by means of grants from NIH. While some of these studies include analyses of foods, others use USDA's standard reference tables adapted and extended to meet their own special needs.

Meat and Poultry Products Regulations

The Food Safety and Inspection Service (FSIS) safeguards the wholesomeness and nutritional quality of meat, poultry, and their products through inspection and analyses. This agency establishes standards, approves labels for meat and poultry products, and monitors the industry for compliance with inspection laws. As well as analysing the nutritional value of meat and poultry products, both domestic and imported, and determining the presence of food additives, it also assesses the nutritional equivalency of imitation and traditional foods and the effects of processing procedures on nutritional quality. For these varied purposes, FSIS conducts laboratory analyses and uses standard reference tables. The Food and Drug Administration (FDA) of the Department of Health and Human Services performs similar regulatory functions for foods other than meat and poultry.

Food Assistance Programmes

The Food and Nutrition Service (FNS) operates 10 major food assistance programmes. Through these programmes, populations at nutritional risk receive food or resources with which to obtain food. Some, such as the Food Stamp Program, provide assistance to the low-income population. Others are designed to help meet the nutritional needs of particular groups, such as children in school and in child care; pregnant and postpartum women, infants, and young children (WIC); and elderly persons. Studies of the effects of programmes on the diets of participants require food composition data, as does the development of standards for programme benefits and compliance.

The legal standard for benefits from the Food Stamp Program is the cost of foods in the thrifty food plan – the least costly of four food plans for nutritious diets developed and costed monthly by HNIS [12]. These food plans reflect food consumption and nutrient levels as reported in the Nationwide Food Consumption Surveys (NFCS), changed only as necessary to meet nutritional goals for 17 food substances. The quantities of foods in the food plans are specified with the assistance of a computerized mathematical model.

Meal patterns for the school lunch and breakfast programmes are designed with the nutrient content of foods they contain as a major consideration. An alternative to using meal patterns as the basis for compliance in the school lunch programme is being field-tested; this involves instead the use of a nutrient standard for the lunch. HNIS has co-operated with FNS in developing software and special data bases for this test. Another programme, WIC, distributes vouchers for foods with nutrient content that will supplement the diets of participants.

NATIONAL NUTRITION MONITORING SYSTEM

As part of the National Nutrition Monitoring System, HNIS maintains the National Nutrient Data Bank, sponsors related food composition research, and conducts surveys in order to monitor the food and nutrient content of American diets. The conversion of information on food consumption into nutrient consumption is the agency's major use of food composition data. The nutrient consumption of the nation's population is monitored at three levels: food in the US food supply, food used by households, and food eaten by individuals. To make these conversions, nutrient data are needed both for food as purchased and food as eaten; and data for each are needed by appropriate units of weight. Representative values – those for the food as purchased or as eaten throughout the country year-round – are appropriate for such conversions.

The food supply data, estimated for each year since 1909, show trends in quantities of nutrients available to the population and in the food sources of these nutrients. For example, the data show that dietary fat in the food supply has increased since the beginning of the century from 124 g to 166 g per person per day [13]. The increase comes primarily from fats and oils. These statistics require that food composition data reflect the foods available for consumption during given periods. For example, the composition of poultry or pork in the 1980s is not the same as it was two decades earlier, and this series reflects those differences.

The food supply covers fewer than 400 basic foods – foods before they are combined into food mixtures. Nutrient data for these relatively few foods in their primary forms are generally readily available. Because of this, estimates of a nutrient's content in the food supply usually can be made before estimates for diets of households and individuals, which contain many more foods. An example is zinc [28]. Several years ago, we had sufficient data to assess the food supply and recognize that the zinc levels of 11.5 to 12.5 mg per person per day since 1909 are not sufficient to provide Americans with the recommended dietary allowance of 15 mg per day [11]. Sufficient data are only now becoming available on enough foods to allow us to assess zinc levels in household and individual diets. Many other countries make nutritional assessments of their food supplies based on statistics such as these.

Information on household food use and individual food intakes is collected by interview about every 10 years in Nationwide Food Consumption Surveys, the most recent of which was conducted in 1977/78. The next is planned for 1987. In 1977/78, data were collected in the 48 conterminous states for about 38,000 individuals in 15,000 households [24]. The surveys showed the kinds and amounts of foods that households used during one-week periods and the food and nutrient intakes of individual household members for three-day periods.

The food used by the household, expressed in pounds of foods as purchased or brought in, is translated into the nutrient content of the household diet by using tables

74

of the nutritive values of the edible parts of the food as purchased. Vitamin content is adjusted for usual losses during cooking. The food intake by individual household members, expressed in grams of foods eaten, is translated into nutrient content by using tables of the nutritive value of the edible portion of foods. To make these calculations requires representative values for each of the thousands of foods in forms that Americans buy for use at home and of foods that they eat both at home and away from home. Analyses are not available to provide reliable data for all nutrients for all of these foods. Therefore, some values are imputed from another form of the food or from a similar food to provide the best estimates of the nutrient content of diets.

Another data file is required for survey data analysis. It converts the quantity of household food reported in packages of rice, loaves of bread, cartons of milk, heads of lettuce, and cans of beans into pounds. It also converts quantities of food individuals ate – a hamburger pattie, a cup of milk, an orange, a piece of cake – into grams of edible food.

Over the decades, these surveys have required data not only on more and more foods but also on more and more nutrients. For example, data on various types of dietary fibre, carotenes, selenium, and tocopherol are now of special interest. Our nutrient data research is directed toward filling these and other needs. In the meantime, diet appraisals based on composition that are limited must be recognized as tentative and possibly misleading. Another concern in appraising diets – and one that can be addressed at this time only in a general way – is the availability of nutrients to the body as affected by the food carrier and other foods and drugs consumed.

As diets are monitored for more nutrients, the descriptive information required about the foods in the survey increases. For example, if information on only total fat in the diet is required, the specific ingredients in margarine are not of major concern. However, if the levels of fatty acids are to be measured, information on the types of oils in the margarine must be obtained. The sodium content of a vegetable depends on whether it was prepared from food in the raw, canned, or frozen, and on what seasonings were added in processing and preparation.

HNIS has considered such problems in planning the questionnaire and analysis for the Continuing Survey of Food Intakes by Individuals begun in April 1985 [20]. New survey questions were introduced in an attempt to obtain the food descriptions needed for more reliable estimates of the levels of cholesterol, total fat, fatty acids, and sodium in diets. Some of this descriptive information the respondent may not be able to provide. The decision must then be made as to whether the missing information can be estimated with reasonable accuracy or whether such estimates are so poor as to be misleading.

Information on food composition has expanded and improved considerably since the last national survey was conducted in 1977/78. At that time, intakes were assessed for food energy; for the macro-nutrients – protein, fat, and carbohydrate; for four mineral elements – calcium, iron, magnesium, and phosphorus; and for seven vitamins – A, thiamine, riboflavin, niacin, B_6, B_{12}, and C.

For the new Continuing Survey, HNIS nutrient data specialists have updated and extended the data file to include 12 more nutrients and other food components than used in the previous survey: three classes of fatty acids, cholesterol, sodium, potassium, zinc, copper, carotenes, vitamin E, folacin, and dietary fibre. This data file will show the content of food energy and some 27 food components for 100-gram portions of each of about 4,000 foods as consumed, and will contain the most up-to-date information in USDA's Nutrient Data Bank. It is important that users of these data and of the survey results based on them recognize that values for these food components are not equal in reliability. For some food components relatively strong data support the values. Data for others – dietary fibre, vitamin E, and carotenes – are less well founded. If analytical data are not available, values are imputed.

The food consumption survey activities discussed are specified in an implementation plan for the National Nutrition Monitoring System sent to Congress by USDA and DHHS in 1981 [15]. The core surveys of the system are USDA's Nationwide Food Consumption Survey (NFCS) and DHHS's National Health and Nutrition Examination Survey (NHANES).

NHANES collects information for one day's food intake and the frequency of consumption of selected foods; its primary focus is on the nutritional health status as determined by physical examination. The dietary data are assessed for nutrient content using composition tables, and the new data file developed for the Continuing Survey will be updated for use in future NHANES. Some analyses of NHANES data attempt to associate the nutrient content of a person's diet with his health status. Such studies might benefit from more precise data on the composition of the food eaten by that person than is given by the representative values in reference tables.

The statistics from NFCS and NHANES tell much about the diets and state of the nutritional health of the population. They make possible the monitoring of the dietary status of the American population to identify populations at risk, problem nutrients, food and eating patterns, and diet determinants. They have shown that many American diets do not meet desired dietary standards [14]: many are short of recommended levels for certain nutrients such as calcium, iron, zinc, and folacin, some are short in vitamins A and C, and many exceed moderate levels of fat, sodium, and added caloric sweeteners.

A systematic exploration of uses of food consumption information was made by the Food and Nutrition Board (FNB) [10]. Its report of June 1984, *National Survey Data of Food Consumption: Uses and Recommendations*, identifies the primary uses of food consumption survey data and recommends effective means of obtaining those data. Many of the uses listed are secondary uses, in that they require information on the nutrient content of diets and the nutrient contribution of specific foods to diets.

Dietary data from these surveys are used for many purposes: to monitor dietary status, to measure the economics of food consumption, to formulate and show the effects of food assistance and regulatory programmes, and to provide the basis for food selection guides – to name but a few.

NUTRITION EDUCATION

USDA has for decades provided information to help Americans improve their food selections – information based on knowledge of human nutritional requirements, food consumption, and food practices [29]. The Dietary Guidelines published jointly by USDA and DHHS in 1980 and revised in 1985 [27] advise Americans to: (a) eat a variety of foods; (b) maintain desirable weight; (c) avoid too much fat, saturated fat, and cholesterol; (d) eat foods with adequate starch and fibre; (e) avoid too much sugar; (f) avoid too much sodium; and (g) if you drink alcoholic beverages, do so in moderation.

Putting these guidelines into practice requires an understanding of the composition of foods. It is important to know the "good" sources of a nutrient – those foods of which a serving provides a substantial part of the day's need – and the "important" sources – those foods that make worthwhile contributions of a nutrient because substantial amounts of the food are consumed [4]. "Important" nutrient sources have been identified by ranking the foods Americans consume by their average contribution of nutrients to the diet.

Eat a Variety of Foods

This is a simple guideline representing a complex food selection technique. What is intended is "Eat the kinds and amounts of foods that will provide the minerals and vitamins your body needs." If Americans are to select diets that provide the recommended amounts of essential nutrients, they must increase their understanding of food composition. To help them, educators must translate complex food composition tables, which were developed for researchers and professionals, into simpler forms [25].

Maintain Ideal Weight

NHANES estimates that one-fourth of all Americans are overweight [1]. Weight-control plans that call for a varied diet made up of low-calorie foods require the use of food composition tables. An USDA pocket calorie guide, *Calories and Weight* [18], gives such advice, as does *Food 2: A Dieter's Guide* [8].

Avoid Too Much Fat, Saturated Fat, and Cholesterol

Fat intakes over three days from our 1977/78 survey were generally above the 30 to 35 per cent of calories suggested by some authoritative groups [5]. For example, among adults of 35 to 50 years of age, only 13 per cent of the males and 17 per cent of the females had fat intakes that provided less then 35 per cent of total calories [14]. Over one-third of the respondents got 45 per cent or more of their calories from fat.

The amount of saturated fat in diets reported in NFCS has not been estimated. Average cholesterol intakes have been roughly estimated using the limited information available on the types of fats and oils consumed. Intakes are highest for adult males and teenage boys, at 400 to 525 mg per day [16].

Efforts to develop guidance that will help Americans avoid too much fat, saturated fat, and cholesterol are complicated by the occurrence of fat and cholesterol in many foods that are important for their contribution of nutrients in US diets. Solutions depend on knowing the fat, cholesterol, and vitamin and mineral content of foods. *Food 3: Eating the Moderate Fat and Cholesterol Way* addresses this issue [9].

Eat Foods with Adequate Starch and Fibre

When the fat in the diet is reduced, calories must come from other sources. This guideline stresses that the source should be foods rich in complex carbohydrates. Food composition tables identify these as dry beans and peas, nuts, seeds, vegetables, fruits, and grain products. Increasing the use of foods that are high in complex carbohydrate will also increase the amount of fibre in the diet. Food composition tables in *Agriculture Handbook No. 8* have provided crude fibre values for some time; but these values are no longer considered useful. The content of *dietary* fibre, now believed to be more meaningful, is not known for many foods. To further complicate the matter, available values may not be comparable because of the lack of standard analytical methods. The new data file for the Continuing Survey provides the most up-to-date values for dietary fibre and imputes values where data are missing. However, the values are not sufficient at this time to estimate the dietary fibre content of American diets with any degree of precision.

Avoid Too Much Sugar

Sugar, as intended in this guideline, is not just sucrose but all caloric sweeteners eaten with foods or as ingredients in foods. The few composition tables that attempt to provide data on added sweeteners should be used with caution, since they are useful only for providing rough estimates of the intakes of added sweeteners in diets [31].

Avoid Too Much Sodium

The sodium content of diets is notably difficult to measure, but there is little doubt that Americans consume more sodium than they need [30]. The amounts of sodium in many commercial products and those added in preparation and at the table are virtually unknown. *Sodium Content of Your Food* is a bulletin that provides sodium values for commonly used foods. *Sodium, Think About It*, published by the Food and Drug Administration and FSIS, groups foods by their sodium content for general guidance in following this dietary recommendation.

Food selection guides since the turn of the century have used food composition data as one of several bases. Atwater in 1894 suggested diets for the American male based on content of protein, carbohydrate, fat, and "mineral matter" [3]. Other guides, including the *Basic Four* in the 1950s [22] and the *Hassle-free Guide* [7], focus on the types of foods that are important sources of vitamins and minerals. The *Hassle-free Guide* also considered calories, sugar, fat, sodium, and fibre. The dietary guidelines published in 1980 [27] have been interpreted by USDA and DHHS at the federal level, by state extension services, by other local groups, and by the food industry. USDA efforts include consumer bulletins, such as *Ideas for Better Eating* [23], *Food 2: A Dieter's Guide*, *Food 3: Eating the Moderate Fat and Cholesterol Way* [9], and *Sodium Content of Your Food* [26].

The HNIS staff co-operated with the American Red Cross in developing a six-session nutrition course based on the dietary guidelines and other timely nutrition messages [17]. This course, "Better Eating for Better Health," has been offered by local Red Cross chapters across the country, starting in 1984. Its food guidance system is a food wheel, with suggested servings from several groups of food. The wheel is supplemented with information on food composition and "trade-offs," or food substitutions – all based on food composition data [2].

The Extension Service – the educational arm of USDA – has nutrition specialists in every state and home economists in almost every county. Their nutrition-related mission is to improve food management skills, the quality of diets, and the nutritional health of the public by transferring research-generated knowledge. They help the public to understand how foods that differ in nutrient content can be combined in diets that are nutritionally sound. Food composition data are fundamental to the development of these educational messages. Many states have developed interactive computer programs for appraising the nutritional quality of diets, and use these programs as educational tools. HNIS and the Extension Service are initiating a joint project for a diet assessment system using data bases that will be updated regularly.

DISCUSSION

This review of food composition data needs illustrates that the uses of food composition data are numerous and that data needs differ.

Nutrients and other food components for which data are needed differ by:
– energy and proximate composition;
– plus selected vitamins and minerals;
– plus other food components of dietary and health concern;
– plus pesticides, residues, toxicants, and additives.

Foods for which data are needed differ by:
– stage in the production-to-consumption chain (food supply – pre-retail; food as purchased; food as eaten);
– amount of processing and preparation (raw carrots; frozen carrots; carrots frozen in butter sauce; carrots cooked from those frozen in butter sauce);
– nature of descriptive characteristics (grapefruit, pink grapefruit, pink grapefruit grown in Texas, pink grapefruit grown in Texas under specified conditions and picked x weeks before market).

Degree of precision and accuracy needed differs:
– specific food must be analysed;
– samples of food must be specific for cultivar, place grown, and other characteristics that affect composition;
– values must be representative of a food as consumed in the United States;
– representative values are required, plus imputed values if analytical data are not available.

Units of measure of food differ:
– values for edible part of a pound of food as purchased;
– values for 100 grams of edible food;
– values for a household measure (cup, slice, piece, etc.) of food.

Complexity of presentation format differs:
– for scientists;
– for educators and health professionals;
– for the general public.

APPROACHES TO MEETING DATA NEEDS

HNIS attempts to meet most of these needs through its publications and computerized data tapes. *Agricultural Handbook No. 8* was published in a single volume in 1963, and is being revised in 23 sections, each dealing with one group of foods [19]. Each section contains tables giving representative values of the edible parts of a pound of food as purchased, or 100-gram portions of edible food, and of food by household measures. These tables cover food in its raw state, food as it is processed for market, and food as it is prepared for consumption. Mixtures as marketed and certain commonly used home-prepared mixtures are included. The tables show the number of samples and the standard error for each value. If analytical data are not available, the tables do not show a value.

Data are shown for moisture, energy, protein, total lipid, carbohydrate, crude fibre, ash, dietary fibre (insoluble), calcium, iron, magnesium, phosphorus, potassium, sodium, copper, manganese, zinc, ascorbic acid, thiamine, riboflavin, niacin, vitamin A, pantothenic acid, vitamin B_6, folacin, vitamin B_{12}, tocopherol, cholesterol, commonly occurring fatty acids, pytosterol, and amino acids. The published sections of *Agriculture Handbook No. 8* present detailed information in a complex format designed to be useful to scientists. Its loose-leaf format allows information for a single food to be updated. Selected data are presented in ways that are more easily used by educators and the general public.

The tables of nutrient data are also available in machine-readable form from the National Technical Information Service, US Department of Commerce. Imputed values for nutrients for which analytical data are not available are given in special data files that represent the staff's best judgement based on analyses of other forms of the food or of similar foods. Such values are flagged on the tape. The Nutrient Data Bank also contains data separately for foods with different characteristics, some of which are not shown separately in the tables. Specific values can be made available on request.

HNIS does not attempt to estimate the levels of pesticides, residues, toxicants, and additives in foods, except those additives that contribute energy and nutrients. The Agency does not have its own laboratories for conducting special-purpose analyses. Certain other parts of USDA, mainly ARS and FSIS, do have such laboratories, which they use to meet their need for highly precise data.

Food composition data have come a long way since the Second World War years when the content of a few nutrients in the civilians' share of the food supply was calculated in USDA on a clanking calculator. This is to the credit of many: the HNIS Nutrient Data Research Branch and their extra-mural analytical programme, the ARS Nutrient Food Composition Laboratory, the food industry, and others who develop sound analytical methods and conduct analyses and share them for use in the Nutrient Data Bank. But there is more to be done. Some HNIS objectives are as follows:
1. Improve nutrient data documentation by implementing an objective system for evaluating and specifying the quality of the data presented.
2. Improve data bases for several food components – dietary fibre, carotenes, tocopherol, selenium, copper, manganese, folacin, chromium, and molybdenum.
3. Keep up with the food production, marketing, and consumption changes in the country by covering new foods and determining the effects of new procedures on nutrient levels.

4. Complete a computerized food formula (recipe) file to develop and document the nutrient content of home-made mixtures and to approximate ingredients in commercial mixtures in the data sets for use in national surveys. This file can be used to assess diets reported in surveys in terms of the basic foodstuffs they contain.
5. Develop guidelines for the proper selection and use of food composition data for a variety of purposes.
6. Encourage users to evaluate carefully the computerized data files they use.
7. Design tools for helping the public to understand that food composition information is useful in selecting nutritious and healthful diets.

REFERENCES

1. S. S. Abraham, M. D. Carroll, M. F. Majjur, and R. Fulwood, *Obese and Overweight Adults in the United States. Vital and Health Statistics*, series 11, no. 230 (US Department of Health and Human Services, Hyattsville, Md., 1983).
2. American National Red Cross and US Department of Agriculture, *Better Eating for Better Health* (American National Red Cross, Washington, D.C., 1984).
3. W. O. Atwater, "Foods: Nutritive Value and Cost," *Farmers Bull.* (USDA, Washington, D.C.), no. 72 (1894).
4. O. M. Batcher and J. M. Nichols, "Identifying Important Food Sources of Nutrients," *J. Nutr. Educ.*, 16: 177 (1984).
5. P. M. Behlen and F. J. Cronin, "Dietary Recommendations for Healthy Americans Summarized," *Food Econ. Rev.*, no. 3 (1985).
6. J. R. Block, "USDA's Commitment to Nutrition in the 80's," *Nutr. Today*, 18(6):6 (1983).
7. C. A. Davis, L. H. Fulton, L. Light, D. D. Odland, L. Page, N. R. Raper, and R. S. Vettel, "Food," *Home and Garden Bull.*, no. 228 (1979).
8. C. A. Davis, L. H. Fulton, L. Light, D. D. Odland, and C. E. Woteki, *Food 2: A Dieter's Guide*, developed by USDA (American Dietetic Association, Chicago, Ill., 1982).
9. C. A. Davis, L. H. Fulton, L. Light, D. D. Odland, and C. E. Woteki, *Food 3: Eating the Moderate Fat and Cholesterol Way*, developed by USDA (American Dietetic Association, Chicago, Ill., 1982).
10. Food and Nutrition Board, *National Survey Data on Food Consumption: Uses and Recommendations* (National Academy of Sciences, Washington, D.C., 1984).
11. Food and Nutrition Board, *Recommended Dietary Allowances*, 9th ed. (National Academy of Sciences, Washington, D.C., 1980).
12. R. L. Kerr, B. B. Peterkin, A. J. Blum, and L. E. Cleveland, "USDA 1983 Thrifty Food Plan," *Food Econ. Rev.*, no. 1: 18 (1983).
13. R. M. Marston and N. R. Raper, "Nutrient Content of the US Food Supply," *Natl. Food Rev.*, no. 29 (1984).
14. E. M. Pao and S. J. Mickle, "Problem Nutrients in the United States," *Food Technol.*, 35(9): 58 (1981)
15. B. B. Peterkin, "National Nutrition Monitoring System," *Family Econ. Rev.*, no. 4: 15 (1984).
16. N. R. Raper, "Cholesterol Intakes of Individuals" unpublished paper (1984).
17. A. M. Shaw and P. M. Marsland, "Better Eating for Better Health," *Family Econ. Rev.*, no. 4: 20 (1984).
18. US Department of Agriculture, "Calories and Weight," *Rev. Agric. Information Bull.*, no. 364 (1981).
19. US Department of Agriculture, "Composition of Foods: Raw, Processed, Prepared," *Agriculture Handbook No. 8* (Science and Education Administration, USDA, Washington, D.C., 1976–1984).
20. US Department of Agriculture, *Continuing Survey of Food Intakes by Individuals*, Report no. 85–1 (USDA, Washington, D.C., 1985).
21. US Department of Agriculture, *Directory: Human Nutrition Activities* (USDA, Washington, D.C., 1984).

22. US Department of Agriculture, *Food for Fitness: A Daily Food Guide*, Leaflet no. 424 (USDA, Washington, D.C., 1958).
23. US Department of Agriculture, *Ideas for Better Eating* (USDA, Washington, D.C., 1981).
24. US Department of Agriculture, *Nationwide Food Consumption Survey 1977–78 Reports* (USDA, Washington, D.C., 1979–1984).
25. US Department of Agriculture, "Nutritive Value of Foods," *Rev. Home and Garden Bull.*, no. 72 (1981).
26. US Department of Agriculture, "Sodium Content of Your Food," *Rev. Home and Garden Bull.*, no. 233 (1980).
27. US Department of Agriculture and US Department of Health and Human Services, "Nutrition and Your Health: Dietary Guidelines for Americans," *Home and Garden Bull.*, no. 232 (1980, 1985).
28. S. O. Welsh and R. M. Marston, "Zinc Levels of the U.S. Food Supply – 1909–1980," *Food Technol.*, 36(1): 70 (1982).
29. I. D. Wolf and B. B. Peterkin, "Dietary Guidelines: The USDA Perspective," *Food Technol.*, 38(7): 80 (1984).
30. C. E. Woteki, N. R. Raper, and H. Riddick, "Sodium Intakes of Individuals," in T. M. Freeman and O. W. Gregg, eds., *Sodium Intake – Dietary Concerns* (American Association of Cereal Chemists, St. Paul, Minn., 1981), p. 105.
31. C. E. Woteki, S. O. Welsh, N. Raper, and R. M. Marston, "Recent Trends and Levels of Dietary Sugars and Other Caloric Sweeteners," in S. Reiser, ed., *Metabolic Effects of Utilizable Dietary Carbohydrates* (Marcel Dekker, New York, 1982), p. 1.

8
Using Food Composition Data to Communicate Nutrition to the Consumer

CAROL T. WINDHAM, NOREEN B. SCHVANEVELDT,
BONITA W. WYSE, and R. GAURTH HANSEN

Department of Nutrition and Food Sciences, Utah State University, Logan, Utah, USA

INTRODUCTION

The relationship of diet and health is a foremost concern of the public. Consumers want to know what to eat to assure nutritional adequacy, health, and well-being. Nutritionists are continually re-evaluating information on food composition, nutrient requirements, food practices, and the relationships between diet, health, and disease as a basis for developing educational programmes for the public.

All of this information (human nutrient requirements, food consumption practices, diet, health, and disease relationships, as well as nutritional guidance) must be underpinned by quality food consumption data. The amount and specificity of nutrient information communicated varies with different audiences, their level of sophistication and whether we are targeting the general population, groups at risk, or specific groups of interest such as students, hospitalized patients, athletes, or individuals. However, the data needed by professionals to support food guidance research and development are fairly specific.

For foods that comprise the majority of calories consumed by the population, extensive nutrient composition data are necessary. It is no longer adequate to have data concerning only the traditional nutrients, i.e. water, kilocalories, protein, total fat, carbohydrate, calcium, phosphorus, iron, vitamin A, thiamine, riboflavin, niacin, and ascorbic acid. Increasingly, the food industry, educators, and consumers need information for such nutrients as magnesium, potassium, sodium, zinc, copper, manganese, selenium, pantothenic acid, total vitamin B_6, folacin, and vitamin B_{12}, as well as for specific fatty acids, cholesterol, specific amino acids, carotenoids, and dietary fibre. Many of these components represent emerging interests, and for this reason the demand for reliable data exceeds availability.

Recognizing the potential of computer technology to help meet these data needs, as well as the analytical needs of nutrition educators, Utah State University and the Human Nutrition Information Center, US Department of Agriculture (USDA) developed NUTREDFO, the Nutrition Education Information System [34], designed for use by professionals as a tool for nutrition guidance research and development of nutrition guidance information.

NUTREDFO SYSTEM DEVELOPMENT

NUTREDFO is a computerized data base system containing three data files and a nutrient analysis software program. Because nutrition education researchers need expansive nutrient information on foods representative of consumption practices, primary consideration in development of the data base was given to: (a) providing complete nutrient profiles of food items with all values verified with original source data; (b) identifying and including commonly consumed foods available in the market-place in simplest edible form; (c) developing analysis and data manipulation programs with maximum flexibility for users; and (d) selecting and validating all data to assure a high-quality, documented data base in which the user could have confidence. The data-base system has been field tested in five university settings with positive results.

The first of the NUTREDFO data files is a permanent file of nutrient values for about 460 commonly used foods. This file is called "permanent" because it cannot be changed in any way by the user. With only a few exceptions, these foods are frequently consumed by the US population and are nationally available for purchase at the retail level. In general, data are given for each food in its most basic, edible state. For each food, values are supplied for 26 nutrients and food constituents (energy, protein, fat, carbohydrate, calcium, iron values, magnesium, phosphorus, potassium, sodium, zinc, thiamine, riboflavin, niacin, vitamins A, B_6, B_{12}, and C, pantothenic acid, folacin, alcohol, added sugar, saturated, mono- and polyunsaturated fats, and cholesterol). All zero values were validated to eliminate incomplete or missing data. Values were imputed in instances where published values were not available.

The second NUTREDFO file is a temporary nutrient file that was created to enable users to temporarily store nutrient values for food items that are not part of the permanent nutrient data base. It can be accessed at any time in the NUTREDFO software program, allowing the user to tailor that program for special project or research needs.

The third file provides documentation on all values in the permanent nutrient file. An 11-digit coding system provides on-line information about the original source for each nutrient and constituent value.

The system capabilities of the nutrient analysis software program are listed in table 1. NUTREDFO will: list nutrient and food constituent levels; rank order foods by nutrient level, per cent standard (1980 Recommended Dietary Allowances (RDA) [10], USRDA, nutrient density standards) or Index of Nutritional Quality [29] for nutrients; calculate means, standard deviations, ranges for nutrients in food groups, days, meals; calculate per cent calories from protein, fat, carbohydrate, alcohol; calculate nutrient amounts per 1,000 kcal for meals, days, multi-day grouping; and allow users to change serving size and nutrient levels of foods to examine the effects for nutrition education purposes.

NUTRIENT AND FOOD CONSTITUENT DATA SOURCES

A major consideration in developing NUTREDFO was to obtain the most accurate nutrient and food constituent values possible. For those nutrients in foods for which

Table 1. NUTREDFO system functions for analytical testing

1. List nutrient and food constituent levels for any food both in the permanent and temporary data file (this function lists any of the 26 nutrients, food constituents, and serving size specified by the user).
2. List foods in rank order by nutrient or food constituent based on contribution of that nutrient or constituent per serving of the food item.
3. List foods in rank order for any specified nutrient based on the percentage of standard for that nutrient. Any one of three standards can be used in this function, including the 1980 RDA [10], Single-value Nutrient Allowances per 1000 Kilocalories [28] or the US Recommended Daily Allowance [69].
4. List foods in rank order for a user-specified nutrient based on its Index of Nutritional Quality (INQ) [29]. This rank-ordering uses the same standards as in those in no. 3 above.
5. Calculate the mean, standard deviation of the mean, and minimum/maximum range for selected nutrients in user-specified groups of foods.
6. Allow the user to temporarily change serving sizes and nutrient and food constituent levels in both the permanent and temporary data files. This function can be used to examine the effects of changes in nutrient levels on selected foods, but does not jeopardize the security of data in either file.
7. Calculate the nutrient value of menus. Menus can be for individual eating occasions, individual days, and for a group of days.
8. Calculate the Index of Nutritional Quality; percentage of standard for nutrient totals, percentage of calories from total protein, carbohydrate, fat, and alcohol; and calculate values for each nutrient and food constituent per 1000 kilocalories for each eating occasion and each day.
9. Calculate the mean, standard deviation, and minimum/maximum range for each nutrient and food constituent in a multi-day grouping of menus.

analytical values existed the objective was to determine the most appropriate original source of data. For nutrient values for which analytical data did not exist, the most legitimate criteria for imputation had to be identified.

USDA was considered the most appropriate original source of data for raw or cooked foods. Data values available in recently revised sections of *Handbook No. 8* [2, 14, 22, 23, 30, 41, 43, 44, 57, 58, 59, 60] were used preferentially, and then provisional food composition tables developed by USDA [13, 25, 26, 45] and values in the 1963 edition of *Handbook No. 8* [71]. All values were verified with the original source. Whenever possible, analytical iron values from *Iron Content of Foods* [15] were used for items not listed in revised *Handbook No. 8* sections.

USDA computerized data tapes [64, 65, 67] were used to supply data for missing values in *Handbook No. 8* printed tables. A recent article by Hepburn [31] discussed these data tapes. In some cases, particularly for vitamins B_6 and B_{12}, published or computerized values had to be recalculated because retention and cooking yield factors had been inconsistently applied [46].

Journal articles by USDA research specialists [3, 4, 5, 6, 9, 16, 17, 18, 19, 20, 21, 37, 38, 47, 53, 55, 56, 72] and other USDA publications [42, 48, 52] were also used as sources for data for specific nutrients and food constituents not available in *Handbook No. 8*, the provisional tables, or the data tapes. These nutrients include fatty acids, cholesterol, zinc, folacin, and sugars in ready-to-eat and granola cereals. Occasionally, food composition sources [8, 49, 70] not published by USDA provided data where none could otherwise be found.

Nutrient values were imputed when acceptable published sources were not available. Imputations were generally derived through mathematical manipulation or adjustment of data from published sources. The preferred approach was to calculate nutrient values for cooked products using data for raw products and applying USDA retention and yield factors. When this was not possible, data values were assumed from similar products, e.g. values for cooked chard were used to estimate values for cooked romaine. An additional recourse was to consult experts in the field, usually Consumer Nutrition Center specialists who were working on particular food and nutrient data and who had access to industry and other data specialists. Calculations for recipe imputations were based on published USDA procedures as used in *Handbook No. 8* [71].

The NUTREDFO system provides specific information about the source for each nutrient and food constituent value. This documentation enables users to have on-line access to the source of every nutrient value in the data base.

FOOD COMPOSITION DATA CHARACTERISTICS AND LIMITATIONS

Many factors influence the reliability and accuracy of nutrient composition data, and an understanding of these factors is essential before working with nutrient data bases. Recently, the strengths and limitations of nutrient data have been addressed in detail in conferences, speeches, and papers [31,51,62]. Of special interest are discussions by Perloff [51] and Hepburn [31] on USDA food composition data. Their papers discuss the uses, strengths, and limitations of food composition data in general. Four issues about food composition data are particularly important: (a) the representativeness of the data; (b) the sources of data; (c) the method used to derive nutrient values; and (d) the adequacy of analytical methodologies.

The representativeness of data depends on the number and quality of laboratories performing food composition analyses, the number and quality of samples used, and any weighting procedures used. Representative data reflect the nutrient composition of food products on a nationwide, year-round basis. Perloff [51] suggests that data are generally more reliable when they are based on analyses of a large number of samples from many locations and when they are compiled from several laboratories.

Often, a weighting scheme is used to average nutrient values from a number of samples. This makes nutrient values more representative of a national food supply and allows varieties of foods that are produced and consumed in larger quantities to be more accurately represented in the final value.

The reliability and usefulness of data also depend on its source. Analytical procedures and methods differ between laboratories and can influence the reliability and accuracy of nutrient information. This underlines the importance of using data compiled from several sources [51].

Published food composition values are derived either through direct analysis or by calculation. When data are obtained from the direct analysis of samples, natural variations in a food sample should be considered. Techniques for analysing samples also change with changes in prevailing cultivars or breeds, food products, and advances in food technology [51].

Values that cannot be averaged from actual analytical data can be calculated using analysed values. For example, protein is calculated from the nitrogen content of the food. Calculations for food mixtures are performed using analytical values for ingredients in the mixture. For cooked foods, calculations are based upon analytical values for the raw products, then adjusted for yield and retention factors.

Calculated values provide a different kind of data than those from direct analysis. The quality of calculated values depends on the quality of the original analytical values and the accuracy of calculation procedures. As advances are made in analytical methodologies, truer measurements of nutrient levels will result in better quality and greater reliability of data from both direct analysis and calculations [51].

The state of analytical methodology is an important factor influencing the availability and quality of food composition data. Hepburn [31] suggests that the status of analytical methodologies is often dependent on the interest expressed by experts in the field as well as on advancements in technology. Interest in certain nutrients shown by the professional community often sparks further research in the development of analytical methods for those nutrients.

INTERRELATIONSHIPS OF NUTRITION EDUCATION AND FOOD COMPOSITION DATA

The NUTREDFO data base was specifically designed for use by professionals as a tool for nutrition guidance research, and for the development of nutrition guidance information and other technical nutrition information. Throughout the world the means by which we have communicated nutrition generally to the population has been a food grouping system [1]. In the United States, the Department of Agriculture has traditionally developed food guides for the purpose of translating dietary allowances into a form which the consuming public can use to improve the nutritional balance of their diets. USDA food guides have evolved over time as a result of an increased understanding of human nutrient needs, food composition, and the relationship of diet to health [32]. All the guides have emphasized that maintenance of good health depends upon consuming a varied diet that will provide adequate amounts of energy and essential nutrients.

The Basic Four Food Groups system, which is currently used in the United States [33], is based upon the balance concept, which assumes that an appropriate mixture of food items from each group will form the foundation of an adequate diet with respect to protein and certain vitamins and minerals for which dietary standards and adequate food composition data were available at the time of the plan's inception. Nutritionists designed the guide to provide approximately 1,200 kcal and at least 80 per cent of the eight nutrients which had Recommended Dietary Allowances (RDA) published in 1953 [39]. These nutrients were protein, vitamin A, thiamine, riboflavin, niacin, vitamin C, calcium, and iron. Because of dietary inadequacies in calcium and vitamins A and C in the American population at that time, the food sources of these nutrients were emphasized – thus the formation of dairy and fruit and vegetable groups. Protein was also cited for specific attention in the meat group because diets containing animal protein sources were expected to contribute micro-nutrients that were difficult to obtain in sufficient amounts from other foods.

This food grouping system assumed that if the need for the key or "leader" nutrients (the basis of the four food groups) were met, then it was likely that requirements for other nutrients, such as vitamins B_6, B_{12}, magnesium, zinc, folic acid, and pantothenic acid would also be satisfied [38]. As new findings are reported on the functions of these other nutrients; as methodologies improve for quantifying these nutrients in foods and biological tissues; and as we examine current food consumption practices, and evaluate and develop diets and menus based on new data, indications are that the assumption behind food commodity groupings and so-called leader nutrients may no longer be valid.

In addition, diseases caused by deficiencies of leader nutrients are currently not the major nutritional concerns in the United States. Other diseases such as cancer, diabetes, hypertension, and heart disease have developed despite the widespread use of the food-group concept in nutrition education. The recent addition by the USDA of a fat, sugar, and alcohol group to the Basic Four addresses these concerns by helping the public become more aware of the dietary levels of nutrients that have been linked to public health problems.

In 1980, USDA and the Department of Health, Education and Welfare (now Health and Human Services) jointly issued recommendations entitled *Dietary Guidelines for Americans* [68]. As with other dietary guidance materials, the guidelines are designed to help consumers make informed choices about foods. The object is to obtain the correct balance of vitamins, minerals, and dietary fibre, without overconsuming salt or calories, especially calories from fat, sugar, and alcohol.

An important companion publication to the *Dietary Guidelines* is *Ideas for Better Eating: Menus and Recipes to Make Use of the Dietary Guidelines* [66]. The menus are designed for healthy adults. There are two versions of each day's menus, one providing 1,600 kcal and the other 2,400 kcal, reflecting the average amounts of energy from foods that women and men, respectively, reported consuming in recent food consumption surveys [12]. Thus, these menus are designed not for weight reduction, but to reflect as closely as possible the recommendations in the *Dietary Guidelines* and RDA. The menus on average contain less than 35 per cent of calories as fat (as suggested in the text of the 1980 RDA), 50 per cent or more calories from carbohydrates, an average cholesterol content lower than the current average consumption, and sodium content within the 1,100 to 3,300 milligrams range recommended in the 1980 RDA publication.

The NUTREDFO data-base system was developed to analyse the menus in *Ideas for Better Eating*. A careful analysis of these menus indicated a number of problems. The protein content, and most vitamin and mineral content, varied between menus, but on average met or exceeded the 1980 RDA for adults. The exceptions were vitamin B_6, folacin, iron, and zinc on the 1,600 kcal diet. Even though these values were low in the menus, they were not as low as usual consumption levels reported at a comparable level of calories in national surveys.

The advances that have been made in improving analytical data for a greater number of nutrients have identified new nutritional problems which, it seems, current nutritional guidance is not addressing or the food supply adequately providing. Nutritional guidelines must reflect current dietary concerns, which in part result from advances in food composition research. The question for nutrition education is, should we change the configuration of the food grouping system to emphasize the nutrients that are now of concern?

During the past few years, several articles have criticized commodity-based food grouping systems [7, 27, 35, 50]. Critics argue that food groups fail to assure nutritional adequacy and are of little relevance to current nutritional thinking. Some have suggested replacing commodity-based systems with nutrient-based systems, thus assuring individuals of a closer approximation to dietary standards [24].

However, commodity food groups have long been used as a basis for nutrition education; they are natural and easily recognized by people with little technical background in nutrition [1, 36]. Nevertheless, the variability of nutrient compositions within a commodity group can lead to inaccurate or misleading nutritional information. A summary of food composition that is based on both commodity groups and similarity in nutrient attributes can be very useful in understanding and explaining the nutritional structure of the food supply.

USING NUTREDFO FOR NUTRITION GUIDANCE RESEARCH

Using NUTREDFO food composition data we applied mathematical clustering algorithms to the classification of foods within commodity groups into subgroups or clusters with similar nutrient compositions [73]. The results obtained depend on the nutrients selected. By using nutrients that have limited availability in the food supply (i.e. vitamin B_6, calcium, iron, magnesium, folacin, and zinc) and those that pose a possible increased health risk (i.e. sugar, fat, cholesterol, and sodium), it is possible to identify quickly those foods that provide adequate amounts of essential nutrients and excessive amounts of nutrients of concern. Furthermore, the clustering algorithm overcomes a problem that has made it difficult in the past to group foods objectively and accurately, namely, that of dealing simultaneously with more than one or two nutrients. As many nutrient attributes as desired can be used and analysed simultaneously by the algorithm.

The algorithm also provides a "cluster centre" or prototype nutrient composition which represents a summary of the nutrients in the foods assigned to a cluster. Figure 1 illustrates clusters for dairy foods: low-fat milks, plain yoghurt, and buttermilk clustered together owing to their high nutrient density and lowest amounts of total and saturated fat and sugar, and low cholesterol contents (fig. 1a); whole milk and natural cheeses grouped together due to their moderate nutrient levels and relatively high fat and sodium contents (fig. 1b); both creamed and low-fat cottage cheese clustered together (fig. 1d), with their high sodium content overriding the differences in fat content of these two products. Subgroups based upon similarities in attributes were also identified in other food commodity groups [73].

The results obtained indicate that this technique provides valuable insight into the nutrient composition of the food supply. Many of the clusters obtained were ones that might have been anticipated. However, some unexpected associations occurred, which, when seen, were quite logical, but would probably not have been predicted. Moreover, in some commodity groups the cluster centres indicate that although the amounts of certain nutrients may vary from one cluster to another they tend to occur in the same proportion. This means that further investigation could lead to a system of "leader" nutrients, those whose presence indicates the presence or absence of other nutrients. The development of the expanded NUTREDFO data base was and will continue to be critical to furthering our reserach in the area of dietary guidance and nutrition education.

Fig. 1. Nutrient density characterization of five dairy group clusters. Food items listed are those most highly associated with each cluster.

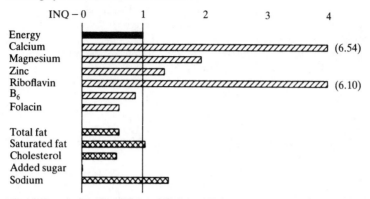

(a) Cluster 1: 1 per cent milk; 2 per cent milk; skim milk; buttermilk; low-fat plain yoghurt.

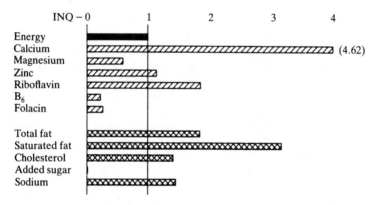

(b) Cluster 2: Whole milk; natural cheese.

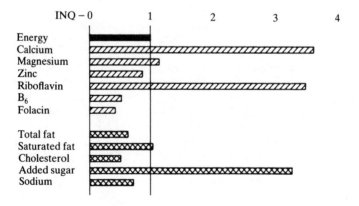

(c) Cluster 3: Chocolate milk, whole and 2 per cent; yoghurt, flavoured and fruited.

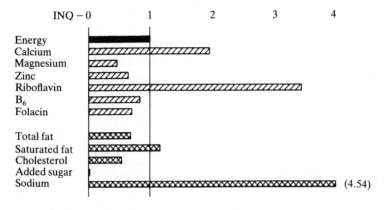

(d) Cluster 4: Creamed cottage cheese; cottage cheese, 1 per cent fat.

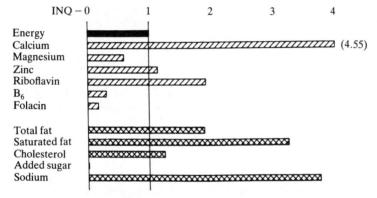

(e) Cluster 5: Processed cheese; Parmesan cheese; blue cheese; cheese spread.

COMMENTS ON SELECTED NUTRIENTS IN NUTREDFO

NUTREDFO users should note that data for some nutrients and food constituents are severely limited or subject to question due to the lack of a standardized methodology. These nutrients are folacin, pantothenic acid, zinc, and "added sugar." Caution should be used when basing nutrition guidance decisions on levels of these nutrients or when reporting these levels in menus, food plans, or other information materials.

Folacin values have been published for only a few forms of various foods. Imputations were necessary to apply these generalized values to specific NUTREDFO foods. In addition, there is conflicting information about the analytical methodology to be used for folacin [62, 63].

Pantothenic acid is also a nutrient for which available methodology provides conflicting information [62, 63]. Although some studies have been conducted to analyse pantothenic acid values, data for processed and prepared foods are still limited [21], and it was necessary to impute levels in many foods.

Published data for zinc are very limited. A 1975 article by Murphy et al. [47] provided data for some important levels of zinc and for widely consumed foods as reported in USDA's 1965/66 Household Food Consumption Survey [11].

Because of the interest in added sugar in the diet, estimates were included in the NUTREDFO system. These values represent the grams of carbohydrate found in excess of those naturally occurring in the food. All "added sugar" values, except ready-to-eat (RTE) cereals, were imputed. RTE breakfast cereals have been analysed for their total sugar content [37, 38]. Since RTE cereals with no added sugar have less than one gram of sugar per 100 grams, total sugar values were used.

Although data on the levels of magnesium, and vitamins B_6 and B_{12}, were reported in the 1977/78 Nationwide Food Consumption Survey (NFCS) [12], data for some foods are limited. In addition there is conflicting information about analytical methodology for vitamin B_{12} [63, 68]. Care should be exercised when interpreting information about these nutrients.

Sodium values in the NUTREDFO permanent data base represent only naturally occurring sodium and sodium added during processing and in recipes. Table salt and salt added during preparation at home are not included. In response to consumer demand for products containing less sodium, some companies are lowering the levels of salt and sodium-containing compounds in products that have traditionally provided substantial amounts of dietary sodium [40, 54, 61]. Consideration will be given to including these products in the NUTREDFO nutrient data file if they become a significant percentage of market-place sales. The NUTREDFO software program has the capability of allowing users to temporarily adjust nutrient values. This capability can be used to test the impact of these new retail products in menu planning or other activities.

RECOMMENDATIONS

As a result of our experiences in developing data bases and putting them to innovative uses, a number of recommendations can be made regarding food composition data and food data systems. We will want more and better data. However, when setting priorities it is important to give first consideration to analysing foods that are representative of current food consumption practices. In addition, it is essential to provide documentation about the source of analytical, imputed, and averaged values to use in resolving discrepancies when they occur. With respect to systems development, in whatever way data are brought together they need to be in a manageable form that is amenable to various types of software. Even a relatively small data base of about 500 foods and 30 nutrients presents difficulties as regards the technical development of software for data use. Something as seemingly straightforward as assigning fields for analysing data can be difficult when systems developers are unaware, for example, of extremes of data values such as vitamin A in liver. While this may seem too simplistic to mention, a great amount and variety of software will be developed for using data in many different ways. Systems developers need to be thinking of fairly sophisticated uses and potential uses for data as they begin to put together a data base.

ACKNOWLEDGEMENTS

Supported in part by USDA Co-operative Agreement 58-3198-2-85 and Utah Agricultural Experiment Station Project 758–3. Journal paper no. 3151 of the Utah

Agricultural Experiment Station, Utah State University, Logan, UT 84322; approved
by the Director.

REFERENCES

1. A. Ahlstrom and L. Rasaneh, "Review of Food Grouping Systems in Nutrition Education," *J. Nutr. Educ.*, 5: 13–17 (1973).
2. B. A. Anderson, "Composition of Foods: Pork Products; Raw, Processed, Prepared," *Agriculture Handbook No. 8–10* (Science and Education Administration, USDA, Washington, D.C., 1983).
3. B. A. Anderson, "Comprehensive Evaluation of Fatty Acids in Foods: VII. Pork Products," *J. Amer. Diet Assoc.*, 69: 44–49 (1976).
4. B. A. Anderson, "Comprehensive Evaluation of Fatty Acids in Foods: XII. Sausages and Luncheon Meats," *J. Amer. Diet. Assoc.*, 72: 48–52 (1978).
5. B. A. Anderson, G. A. Fristrom, and J. L. Weihrauch, "Comprehensive Evaluation of Fatty Acids in Foods: X. Lamb and Veal," *J. Amer. Diet Assoc.*, 71: 412–415 (1977).
6. B. A. Anderson, J. E. Kinsella, and B. K. Watt, "Comprehensive Evaluation of Fatty Acids in Foods: II. Beef Products," *J. Amer. Diet. Assoc.*, 67: 35–41 (1975).
7. I. E. Andersen, "The Pyramid – A New Food Guide," *J. Can. Diet. Assoc.*, 38: 109–110 (1977).
8. W. S. Arbuckle, *Ice Cream*, 3rd ed. (AVI Publishing Co., Westport, Conn., 1977).
9. C. A. Brignoli, J. E. Kinsella, and J. L. Weihrauch, "Comprehensive Evaluation of Fatty Acids in Foods: V. Unhydrogenated Fats and Oils," *J. Amer. Diet. Assoc.*, 68: 224–229 (1976).
10. Committee on Dietary Allowances, Food and Nutrition Board, Commission on Life Sciences, National Research Council, *Recommended Dietary Allowances*, 9th ed. (National Academy Press, Washington, D.C., 1980).
11. Consumer and Food Economics Research Division, Agricultural Research Service, *Food Consumption of Households in the United States, Spring 1965*, Report no. 1 (USDA, Washington, D.C., 1968).
12. Consumer Nutrition Division, Human Nutrition Information Service, *Nutrient Intakes: Individuals in 48 States, Year 1977–78*, Nationwide Food Consumption Survey, 1977/78, Report no. 1–2 (USDA, Hyattsville, Md., 1984).
13. R. Cutrufelli and R. H. Matthews, *Provisional Table on the Nutrient Content of Beverages* (Consumer Nutrition Center, USDA, Hyattsville, Md., 1981).
14. J. S. Douglass, R. H. Matthews, and F. N. Hepburn, "Composition of Foods: Breakfast Cereals; Raw, Processed, Prepared," *Agriculture Handbook No. 8–8* (Science and Education Administration, USDA, Washington, D.C., 1982).
15. J. Exler, *Iron Content of Foods* (US Government Printing Office, Washington, D.C., 1983).
16. J. Exler, R. Avena, and J. L. Weihrauch, "Comprehensive Evaluation of Fatty Acids in Foods: XI. Leguminous Seeds," *J. Amer, Diet Assoc.*, 71: 412–415 (1977).
17. J. Exler and J. L. Weihrauch, "Comprehensive Evaluation of Fatty Acids in Foods: VIII. Finfish," *J. Amer. Diet. Assoc.*, 69: 243–248 (1976).
18. J. Exler and J. L. Weihrauch, "Comprehensive Evaluation of Fatty Acids in Foods: XII. Shellfish," *J. Amer. Diet. Assoc.*, 71: 518–521 (1977).
19. R. M. Feeley, P. E. Criner, and B. K. Watt, "Cholesterol Content of Foods," *J. Amer. Diet. Assoc.*, 61: 134–149 (1972).
20. G. A. Fristrom, B. C. Stewart, J. L. Weihrauch, and L. P. Posati, "Comprehensive Evaluation of Fatty Acids in Foods: IV. Nuts, Peanuts, and Soups," *J. Amer. Diet. Assoc.*, 67: 351–355 (1975).
21. G. A. Fristrom and J. L. Weihrauch, "Comprehensive Evaluation of Fatty Acids in Foods: IX. Fowl," *J. Amer. Diet. Assoc.*, 69: 517–522 (1976).
22. S. E. Gebhardt, R. Cutrufelli, and R. H. Matthews, "Composition of Foods: Baby Foods; Raw, Processed, Prepared," *Agriculture Handbook No. 8–3* (Science and Education Administration, USDA, Washington, D. C., 1978).
23. S. E. Gebhardt, R. Cutrufelli, and R. H. Matthews, "Composition of Foods: Fruits and Fruit Juices; Raw, Processed, Prepared," *Agriculture Handbook No. 8–9* (Science and

Education Administration, USDA, Washington, D.C., 1982).
24. A. H. Gillespie and C. E. Roderuck, "A Nutrient Guide: An Alternative Guide to Food Selection," *J. Can. Diet. Assoc.*, 45: 130–138 (1984).
25. M. S. Goddard, D. B. Haytowitz, and R. H. Matthews, *Provisional Table on the Nutrient Content of Canned Vegetables and Vegetable Products* (Consumer and Food Economics Institute, USDA, Hyattsville, Md., 1979).
26. M. S. Goddard, D. B. Haytowitz, and R. H. Matthews, *Provisional Table on the Nutrient Content of Frozen Vegetables* (Consumer and Food Economics Institute, USDA, Hyattsville, Md., 1979).
27. H. A. Guthrie and J. C. Scheer, "Nutritional Adequacy of Self Selected Diets that Satisfy the Four Food Groups Guide," *J. Nutr. Educ.*, 13: 46–49 (1981).
28. R. G. Hansen and B. W. Wyse, "Expression of Nutrient Allowances per 1000 Kilocalories," *J. Amer. Diet. Assoc.*, 76: 223–227 (1980).
29. R. G. Hansen, B. W. Wyse, and A. W. Sorenson, *Nutritional Quality Index of Foods* (AVI Publishing Co., Inc., Westport, Conn, 1979).
30. D. B. Haytowitz and R. H. Matthews, "Composition of Foods: Vegetables and Vegetable Products; Raw, Processed, Prepared," *Agriculture Handbook No. 8–11* (Science and Education Administration, USDA, Washington, D.C., 1984).
31. F. N. Hepburn, "The USDA National Nutrient Data Bank," *A.J.C.N.*, 35: 1297–1301 (1982).
32. A. A. Hertzler and H. L. Anderson, "Food Guides in the United States," *J. Amer. Diet. Assoc.*, 64: 19–28 (1974).
33. Human Nutrition Center, "Food: A Publication on Food and Nutrition by the US Department of Agriculture", *Home and Garden Bull.*, no. 228 (US Government Printing Office, Washington, D.C., 1980).
34. Human Nutrition Information Service, *Manual for NUTREDO*, Test (1983–1984), Draft (KBL Group, Inc., and Nutritional Guidance and Education Research Division, USDA, Washington, D. C., 1983).
35. J. C. King, S. H. Cohenour, C. G. Corruccini, and P. Schneeman, "Evaluation and Modification of the Basic Four Food Guide," *J. Nutr. Educ.*, 10: 27–29 (1978).
36. P. A. Lachance, "A Suggestion on Food Guides and Dietary Guidelines," *J. Nutr. Educ.*, 13: 56 (1981).
37. B. W. Li and P. J. Schuhmann, "Gas Chromatographic Analysis of Sugars in Granola Cereals," *J. Food Sci.*, 46: 425–427 (1981).
38. B. W. Li and P. J. Schuhmann, "Gas–Liquid Chromatographic Analysis of Sugars in Ready-to-eat Breakfast Cereals," *J. Food Sci.*, 45: 138–141 (1980).
39. L. Light and F. J. Cronin, "Food Guidance Revisited," *J. Nutr. Educ.*, 13: 57–62 (1981).
40. J. Lufty and S. Ferguson, "SS Pierce Launches 'Season by Choice' for Libby's Natural Pack and Vegetables," Consumer Education Program, Rand Public Relations, New York (press release at Libbys Natural Pack Press Luncheon, New York, 23 September, 1982).
41. A. C. Marsh, "Composition of Foods: Soups, Sauces and Gravies; Raw, Processed, Prepared," *Agriculture Handbook No. 8–6* (Science and Education Administration, USDA, Washington, D.C., 1980).
42. A. C. Marsh, R. N. Klippstein, and S. D. Kaplan, "Sodium Content of Your Food", *Home and Garden Bull.*, no. 233 (US Government Printing Office, Washington, D.C., 1980).
43. A. C. Marsh, M. K. Moss, and E. W. Murphy, "Composition of Foods: Spices and Herbs; Raw, Processed, Prepared," *Agriculture Handbook No. 8–2* (Science and Education Administration, USDA, Washington, D.C., 1977).
44. M. A. McCarthy and R. H. Matthews, "Composition of Foods: Nut and Seed Products; Raw, Processed, Prepared," *Agriculture Handbook No. 8–12* (Science and Education Administration, USDA, Washington, D.C., 1984).
45. C. McQuilkin and R. H. Matthews, *Provisional Table on the Nutrient Content of Bakery Foods and Related Items* (Consumer Nutrition Center, USDA, Hyattsville, Md., 1981).
46. A. L. Merrill, C. F. Adams, and L. J. Fincher, "Composition of Foods . . . Raw, Processed, Prepared," *Procedures for Calculating Nutritive Values of Home-prepared Foods: As Used in Agriculture Handbook No. 8*, rev. 1963 (USDA, Washington, D. C., 1966) (ARS 62-13).
47. E. W. Murphy, B. W. Willis, and B. K. Watt, "Provisional Tables on the Zinc Content of Foods," *J. Amer. Diet. Assoc.*, 66: 345–355 (1975).

48. M. L. Orr, *Pantothenic Acid, Vitamin B$_6$ and Vitamin B$_{12}$ in Foods*, Home Econ. Res. Rpt. 36 (USDA, Washington, D.C., 1969).
49. A. A. Paul and D. A. T. Southgate, *McCance and Widdowson's: The Composition of Foods* (HMSO, London, 1978).
50. J. A. T. Pennington, "Considerations of a New Food Guide," *J. Nutr. Educ.*, 13: 53–55 (1981).
51. B. P. Perloff, "Important Information for Nutrient Data Base Users," American Dietetic Association, 65th Annual Meeting, San Antonio, Texas, 20 October 1982.
52. B. P. Perloff, *Supplement to Provisional Table on the Folacin Content of Foods* (Consumer and Food Economics Division, Science and Education Administration, USDA, Hyattsville, Md., 1977).
53. B. P. Perloff and R. R. Butrum, "Folacin in Selected Foods," *J. Amer. Diet. Assoc.*, 70: 161–172 (1977).
54. "Pinching the Salt," *Newsweek*, 23 August 1982, p. 52.
55. L. P. Posati, J. E. Kinsella, and B. K. Watt, "Comprehensive Evaluation of Fatty Acids in Foods: I. Dairy Products," *J. Amer. Diet. Assoc.*, 66: 482–488 (1975).
56. L. P. Posati, J. E. Kinsella, and B. K. Watt, "Comprehensive Evaluation of Fatty Acids in Foods: III. Eggs and Egg Products," *J. Amer. Diet. Assoc.*, 67: 111–115 (1975).
57. L. P. Posati and M. L. Orr, "Composition of Foods: Dairy and Egg Products; Raw, Processed, Prepared," *Agriculture Handbook No. 8–1* (Science and Education Administration, USDA, Washington, D.C., 1976).
58. L. P. Posati and M. L. Orr, "Composition of Foods: Poultry Products; Raw, Processed, Prepared," *Agriculture Handbook No. 8–5* (Science and Education Administration, USDA, Washington, D.C., 1979).
59. J. B. Reeves III and J. L. Weihrauch, "Composition of Foods: Fats and Oils; Raw, Processed, Prepared," *Agriculture Handbook No. 8–4* (Science and Education Administration, USDA, Washington, D.C., 1979).
60. M. Richardson, L. P. Posati, and B. A. Anderson, "Composition of Foods: Sausages and Luncheon Meats; Raw, Processed, Prepared," *Agriculture Handbook No. 8–7* (Science and Education Administration, USDA, Washington, D.C., 1980).
61. Safeway Stores, Inc., *Newsservice*, 12 August 1982 (Safeway Stores, Inc., Washington, D.C. Division, Landover, Md).
62. K. K. Stewart, adapted from a presentation given at the Seventh National Nutrient Data Bank Conference, Philadelphia, Pa., 3–5 May 1982, and personal communication.
63. K. K. Stewart, "The State of Food Composition Data: An Overview with Some Suggestions," *Food Nutr. Bull.*, 5: 53–68 (1983).
64. US Department of Agriculture, *Expansion of Data Published in Nutritive Value of American Foods* (US Department of Commerce National Technical Information Service, Springfield, Va., 1977) (USDA, Agriculture Handbook 456, data set 456–3, no accession no.).
65. US Department of Agriculture, *Expansion of Data Published in Nutritive Value of American Foods* (US Department of Commerce National Technical Information Service, Springfield, Va., 1982) (USDA, Agriculture Handbook 456, data set 456–3, release 3, accession no. PB82-183781).
66. US Department of Agriculture, *Ideas for Better Eating: Menus and Recipes to Make Use of the Dietary Guidelines* (Science and Education Administration, USDA, Washington, D.C., 1981).
67. US Department of Agriculture, *Nutritive Values of Foods, Used in the Nationwide Food Consumption Survey, Basic Individual* (US Department of Commerce National Technical Information Service, Springfield, Va., 1977–1978) (Accession no. PB81-197403).
68. US Department of Agriculture and US Department of Health, Education and Welfare, "Nutrition and Your Health: Dietary Guidelines for Americans," *Home and Garden Bull.*, no. 232 (US Government Printing Office, Washington, D.C., 1981).
69. US Food and Drug Administration, US Department of Health, Education, and Welfare, "Nutrition Labeling," *Federal Register*, 38(49): 6951–6961 (1973).
70. J. H. Walsh, B. W. Wyse, and R. G. Hansen, "Pantothenic Acid Content of 75 Processed and Cooked Foods," *J. Amer. Diet. Assoc.*, 78: 140–144 (1981).

71. B. K. Watt and A. L. Merrill, "Composition of Foods: Raw, Processed, Prepared," *Agriculture Handbook No. 8* (Science and Education Administration, USDA, Washington, D.C., 1975).
72. J. L. Weihrauch, J. E. Kinsella, and B. K. Watt, "Comprehensive Evaluation of Fatty Acids in Foods: VI. Cereal Products," *J. Amer. Diet. Assoc.*, 68: 335–340 (1976).
73. C. T. Windham, M. P. Windham, B. W. Wyse, and R. G. Hansen, "Cluster Analysis to Improve Food Classification within Commodity Groups," *J. Am. Diet. Assoc.*, 85: 1306–1314 (1985).

9
Nutrient Composition Data Uses and Needs of Food Companies

FRED H. STEINKE

Nutritional Biochemistry Department, Ralston Purina Company,
St. Louis, Missouri, USA

INTRODUCTION

The need for good nutrient composition data is recognized by most reputable food companies. The basic objective of any food company is to supply to consumers a highly acceptable food product that will satisfy their nutritional needs while respecting safety standards.

The need for food composition data varies depending on the degree of processing and added value that is applied to the food product. The food industry is a heterogeneous business in that its products range from foods with no (or minimal) processing to highly refined and specific medical foods. These different levels of processing and market orientation determine the need and uses for nutrient composition by food companies.

The frequency of consumption of a food product also determines its importance in the diet and thus the need for nutritional evaluation. Some food products are used only occasionally and therefore have minimal nutritional impact on humans. Others are used regularly and their nutritional composition can be a critical concern for both consumer and manufacturer. The development of complete meals and specialty products for specific groups within the population has further emphasized the need for precise nutritional data.

AVAILABLE FOOD COMPOSITION DATA

The food composition data available are limited to a few sources. An important factor in selecting nutrient composition data is the acceptability of the source to regulatory agencies and scientific groups. The selection of these data is based on well-recognized and scientifically sound sources that will not be subjected to challenge once a product is marketed. This increases acceptance of the products by consumers who are familiar with the need to validate nutrient content for menu planning, for example in hospitals, schools, and other institutions.

USDA, "Composition of Foods," Agricultural Handbook No. 8 and 8–1 to 8–11

This is the primary source of food composition data used in the calculation and formulation of processed food products in the United States [3], and has the advantage of being generally recognized as acceptable to regulatory agencies (USDA, FDA, FTC, etc.), and nutrition and dietetic communities. However, it must be recognized that much of the data in *Handbook No. 8* is 20 to 30 years old, incomplete, and may not apply to current food products, for example to basic staples which have changed due to genetics, and processed foods which are continually undergoing development and change to meet consumer demands. *Handbook No. 8* values can be used in international trade, but often are not acceptable for both scientific and political reasons.

Other Food Composition Data

The FAO Nutritional Studies, no. 24, which gives amino-acid data on foods [1], is widely accepted in developing and socialist countries. The data compiled in the publication are considerably older and focus prominently on basic food ingredients; they are not the same as those given by USDA. McCance and Widdowson's *The Composition of Foods* [2] provides another excellent source of nutrient composition data, which, while it is directed primarily at foods in the United Kingdom, is relatively current and covers a wide range of nutrients.

Most food companies generate nutrient composition data on their own products, on specific ingredients and on competitive products, and many are particularly well equipped for conducting food composition analyses. Since these analyses are kept current through constant monitoring programmes, they are often more reliable than existing data-base information. Companies may supply data on their products for use in data bases, but may not make available other ingredient information. Similarly, some proprietary information on products may not be released, especially if the products are subject to periodic changes in formulation.

Regulations require that a high percentage of a product meets or exceeds the nutrient levels stated on the label. Label information may therefore understate nutrient composition in order to comply with regulatory needs, with the result that average or typical values are often higher than specified.

Manufacturers' product data sheets are sometimes used when information from a recognized data base is either unavailable or inappropriate. The latter may be due to product changes or to improvements in analytical procedures. Most manufacturers have nutrient data available on their food products.

Published data on specific foods or food ingredients are used to supply specific nutrient information when: (a) these are unavailable from documented data bases, (b) the data are new, or (c) the data are more current and reliable than those contained in the data bases. It requires some effort to determine the reliability of the data source and the compatibility of the methodology with similar values from other data sources.

USES OF FOOD COMPOSITION DATA

The need for and use of food composition data by food companies varies according to the type of food product. A great influence on these needs and uses is the type of market in which the products are sold. Those that are advertised as nutritional products or are aimed at nutrition-oriented markets will have a greater need for nutrient composition data than commodity-type products; they will require nutrition support data to validate claims and demonstrate nutritional superiority. Nutrition data may, however, add little or nothing to a company's ability to sell their product and, therefore, be seen as a minor concern. The degree of control that a food company exercises over the composition will influence the need for and use of food composition data. The ability to use nutrient data to increase the sales and acceptance of a product is important to its use in any open market system.

New Processed Food Product Development

Food composition data in ingredients is used to identify the nutrient content of food products under development. The objective may be simply to understand their nutrient composition, or to identify differences from existing products.

The data are used to formulate products to meet specific nutrient composition needs during product development. These types of products are usually designed for specific population segments and nutritional requirements.

Nutrient composition data is also used to identify nutrients that may need to be added to a food product. Nutrient fortification may be targeted to achieve a specific nutrient content or to match the nutrients of an existing competitive food. Final product composition will be confirmed and adjusted after analysis of the product.

Medical and Pharmaceutical Foods

Medical and pharmaceutical types of foods include infant formulas, nutritional supplements, ethical feeding products, and prepared foods with defined compositions for institutional feeding. Special diet preparations which aid in the treatment of specific diseases also require extensive use of food composition data.

This area of food product use and development is more critically based on nutrient composition than that of many other food products. Complete meals or food supplements require accurate and complete data on nutrient content. Menu planning in hospitals and institutions requires complete nutrient data, and the menus must be fully evaluated since the patients are limited in their food selection. The final nutrient composition of the products is usually based on actual analyses rather than calculated values.

A number of sepcific weight-control and weight-loss food products use food composition data to ensure nutrient adequacy. These types of products and programmes increase the need for accurate nutrient information, since the number of foods consumed is limited and can easily result in nutrient deficiencies.

NEEDS AND CONCERNS

The clear identification of food ingredients and subclasses within a food classification is the first and most essential function of any food composition data base. This will allow the food manufacturer to select correctly the food ingredients, or the nutrient composition for the ingredients, used in manufacturing. This is a particular concern with international programmes since there may be less control over foods and ingredients as regards handling, consistency of production, processing, and identification.

Formulated foods may vary in ingredient composition and, therefore, in nutrient composition. Differentiation by ingredient composition is a factor to take into account in developing a data base. The same products in various parts of the world may be identified differently. Since many local names may not completely identify the product, some means of identification needs to be established, such as ingredient composition or recipe. Since processed foods may be modified or changed over a period of time, a programme for elimination, replacement, and updating of compositions should be incorporated into the system.

Validation and correlation of analytical methods is another potential area for major error in nutrient composition. The users need to know the degree of reliability for the values reported. In order to make product comparisons, it is essential that the data be equivalent, if not the same.

Provisions need to be made to allow for rapid searching and identifying of products. If the search procedures are too time-consuming, it will reduce use and acceptance of the system. Compatibility with existing equipment and methods for easy transfer between data systems is also of practical importance.

Updating on a continuing basis will make the data more significant and usable for the food manufacturer, given the high degree of sophistication in this area. New products, processes, and packaging methods are continually being introduced into the food system, and all of these can and do change the composition of food products introduced to the consumer. To accurately reflect nutrient intake, changes in the food systems must be quickly incorporated into the data base.

In particular, better identification of the fibre components of foods and of methods for their analysis is essential. Fibre types may play a significant role in human nutrition and health. This requires a clear definition of food products to allow evaluation of fibre subclasses relative to metabolic effects and interaction with other nutrients in the diet.

SUMMARY

Food manufacturers have uses for nutrient data bases in product development, marketing, and identification of product attributes to the consumer. The food manufacturer needs accurate and current nutrition information on foods that are recognized as scientifically sound by regulatory agencies. The continuing interest of the consumer in nutrition will focus the attention of food manufacturers on nutrient composition and the availability of nutritional data sources.

REFERENCES

1. *Amino Acid Content of Foods and Biological Data on Proteins*, FAO Nutritional Studies, no. 24 (FAO/UN, Rome, 1970).
2. A. A. Paul and D. A. T. Southgate, *McCance and Widdowson's: The Composition of Foods*, 4th ed. (HMSO, London, 1978).
3. US Department of Agriculture, "Composition of Foods: Raw, Processed, Prepared," *Agriculture Handbook No. 8* (Science and Education Administration, USDA, Washington, D.C., 1976–1984).

Managing Food Composition Data

10
Concerns of Users of Nutrient Data Bases

LORETTA W. HOOVER

University of Missouri–Columbia, Columbia, Missouri, USA

INTRODUCTION

Numerous uses of nutrient data bases have been identified in the professional literature during the past twenty years [5, 6, 12]. Users of nutrient analysis software vary greatly in their degree of sophistication and hold widely varying ideas about what a nutrient data base should contain and the features of the associated computer programs. Professionals are seeking reliable data and systems but have no definitive measures for identifying such systems.

A two-tier system of users, with differing needs, has evolved. The first tier interacts directly with depositories of nutrient data – USDA, food manufacturers, and other sources of nutrient data. This group of users includes researchers, clinical practitioners in complex organizations with data-processing support, vendors of nutrient analysis software systems and services, and some private practitioners. These experienced and knowledgeable users who have maintained nutrient data bases for several years are acquainted with the issues of accessibility, installation and updating effort, availability, and computational concerns; they are concerned with the existence of software and data bases that are questionable with respect to accuracy of data, computational results, and dietary guidance.

The second tier of users consists of those individuals who acquire software packages or analysis services from vendors or through resource-sharing arrangements. These are often first-time users who may be unaware of methodological and structural issues. They are often concerned about compatability with a given brand of hardware, cost, and how to select a suitable system.

The first tier of more sophisticated users is likely to benefit most from a data network such as the one envisioned by INFOODS. The second tier will benefit indirectly, as more adequate and comprehensive data bases and software packages become available in the market-place. Meeting the needs of the first tier of users should be given priority.

ACCESSIBILITY

Accessibility factors of concern include source, cost, and timeliness. These factors are important both at the time of initial acquisition of a data base and when an existing data base is being updated.

Maintaining a data base involves much redundant effort on the part of the first tier of users. Since no single source is available for all categories of data, many developers are involved in securing data about commercial products, ethnic foods, and regional specialities from the professional literature, food manufacturers, and unpublished sources. The aggregate time and cost spent on these activities in addition to the periodic costs associated with acquisition of machine-readable data constitute a significant expense for most users in the first tier. Minimizing the total costs for acquiring and maintaining a data base is a concern, particularly for those who do not market their data bases or services.

Timely delivery of data is another aspect of accessibility. Often users have an immediate need for data on specific nutrients or for certain foods. For example, recipes cannot be coded when data are lacking for certain ingredients. Meeting deadlines on research projects may be difficult when data are not readily available.

Knowledge of the availability of new data is also essential. Notification systems are needed to alert users when new data are released. Since some users may wish to access only a portion of the new data, opportunities for extracting selected data records would permit users to maintain up-to-date data bases tailored to their special needs.

With on-demand access, many users would be able to minimize the size of the data base maintained locally, since one could retrieve data from a central depository with the nutrient profile reflecting the most reliable values for each constituent. Acquiring data on a "just-in-time" basis would permit users to avoid data maintenance responsibilities for some data records until a need arises.

INSTALLATION AND UPDATING EFFORTS

The effort required to install and maintain a nutrient data base depends on the specific characteristics of the installation. Because of the effort required when a data record format is changed, a standard format that can remain stable for an extended period of time is desirable. Each time the data record format from a depository is modified, users must either recreate their data bases and reformat data of local interest prior to merging with new data or reformat the new data to be compatible with their existing data-base design.

Many data-base developers utilize nutrient profiles for ingredients to estimate the nutrients in mixed dishes. Information about the quantity of each ingredient is stored in recipe data bases along with cross-referencing codes used to retrieve the nutrient profiles from a nutrient data base when the nutrients per portion are calculated. These recipe data bases facilitate the recalculation of nutrient profiles for mixed dishes by computer. The recalculation process can be involved at different times: periodically, such as monthly; whenever the contents of the nutrient data base are changed; whenever a change is made in the recipe formulation; or when a dietary record is processed with a mixed dish coded as a consumed item. With the maturation of computerized systems for food production and patient care, the use of recipe data bases is more prevalent, particularly in health-care organizations. For these users, the overhead associated with a data record format change is multiplied, since a recipe data base may also require recreation and associated software must be modified to process the new data structures.

Since conversion is expensive and time-consuming, some users may need the opportunity to acquire new data without being required to change to a new data format. Users with minimal resources or limited technical support must be able to maintain compatibility with existing software.

Data should be available in machine-readable form. Manual data entry is time-consuming and error-prone; the probability of locating mistakes is low. Verified data from a reputable source in machine-readable form helps to assure the integrity of nutrient data bases.

Since so many data bases are updated with nutrient profiles for brand-name products and fast foods, a clearing-house or depository for brand-name product information is needed. These data should also be distributed in machine-readable form. The numbering scheme for these products should be co-ordinated with the coding scheme adopted by depositories providing nutrient profiles for generic foods. Also, the measuring units for the amount of food and food constituents should be consistent with those available for generic foods.

A universal or standardized code would be useful to some users to facilitate data-base updating and inter-data-base communications. In a survey of data-base developers conducted in March 1984 by the Data Base Committee of the Ninth National Nutrient Data Bank Conference, 28 of a total of 52 respondents preferred a standardized code or vocabulary.

Recoding within an existing system when the keys to data records are changed is a major task. Cross-reference or linking files are needed to facilitate recoding in other associated data bases such as recipe files. These linking files, which contain the new code number paired with the former code number for each food item, should be available in machine-readable form.

Another practice which would benefit end-users when a new coding system is adopted is application of the new coding system to all food items in an existing data base at one time. Even though new data may not be available at a given time for all entries in the data base, this approach would permit users to accomplish conversion to the new coding scheme and data structure as a single task rather than having to repeat the recoding task each time new data are released. However, if new data are not provided for some of the food items or some items are deleted, users should be informed so that they can replace the obsolete codes with new ones in associated data bases, such as recipe data bases.

Availability of data from a primary depository in user-specified formats would permit customized data bases to be downloaded so that minimal technical expertise would be required of the end-user. On-line retrieval is desirable when the amount of data can be accommodated by a user's equipment and data-transfer costs are economical.

Maintaining compatibility with existing software and other computerized systems reduces the overheads associated with installing new data bases or updating existing ones. Many nutrient data-base users in health-care organizations must compete with other users in their facility for support from the data-processing department. Efficient use of those resources is necessary when data-base conversion is required. Thus, multiple optional arrangements for data transfer would allow users to avoid some installation and updating costs.

DATA AVAILABILITY

The numbers of foods and food constituents per food desired in a nutrient data base vary according to the information requirements of users. Those engaged in research are usually interested in a high level of specificity in food description for those foods of interest to them. In contrast, users in other settings are generally satisfied to use data bases where the items are described in less detail.

In the survey of data-base developers, data on brand-name foods were requested for cereals, candies, fast foods, frozen entrées, margarines, formulated "recipe" items, and fortified foods. Thirty-eight of 52 respondents indicated a need for brand-name data. Some users requested that values be given for all nutrients where Recommended Dietary Allowances (RDAs) have been established. Often, users have little basis for imputing missing values in the nutrient profiles. The nutrient profiles provided in conjunction with nutritional labelling often lack data for constituents of interest to nutritionists.

More complete data are needed for several food constituents. Of particular concern are data on fibre, individual sugars, other carbohydrate fractions, and trace elements. Some developers are also seeking data for the caffeine and alcohol content of foods. The prevalence of missing data in many data bases is a problem. Users are seeking complete profiles for a broad range of nutrients for foods commonly consumed.

COMPUTATIONAL CONCERNS

As indicated above, many nutrient data-base users estimate nutrient profiles for mixed dishes based on institutional or family recipes. Estimates for some nutrients for recipes are included in many cookbooks. Both food-service management systems and nutrient-analysis systems are being designed to facilitate the calculation process. To focus attention on some problems associated with this practice of nutrient estimation, Hoover and Perloff [8] included a simple recipe for a tuna noodle casserole as a computational task in a methodology for assessing computer software.

Several methods are being used to estimate nutrient profiles for recipes, with each requiring associated data not usually present in nutrient data bases. In USDA *Handbook No. 8* [1], nutrient-retention factors, ingredient-weight adjustment factors, and nutrient profiles for raw ingredients have been used to calculate nutrients for recipes. Additional information about ingredient yields has been provided by USDA in another publication, *Handbook No. 102* [10]. Although provisional nutrient retention information has been made available [11], more information is needed for more foods and preparation methods.

In the 1960s, a different calculation method was implemented in food-service software systems [2, 3]. The major difference in calculation method was use of nutrient profiles for the finished form of each ingredient rather than the nutrient profiles for raw ingredients and nutrient-retention factors. Numerous software systems are now using this yield-factor method.

106

Marsh [9] has described a study that focuses attention on the calculation of nutrients for mixed dishes. An approach for calculating nutrients based on "dish retention" was compared with the two methods mentioned above. Although all three calculation methods were used in the study, none of the methods was identified as best in a preliminary discussion of the findings.

Although elaborate procedures can be used to estimate nutrient retention and ingredient yields in a finished product, the actual nutrient profile for a recipe is not known unless a laboratory analysis is performed. Constituent over- or underestimation could adversely impact dietary guidance or menu planning. Further information is needed to identify the best methodology for calculating nutrients for recipes. Without a standard methodology, the results from various nutrient-analysis systems are not likely to be comparable. Depending on the method identified as most reliable to support the calculation process, additional ingredient information will be required. Some of this ingredient information, such as yield factors, may be suitable for incorporation into nutrient data bases.

DATA-BASE AND SOFTWARE PRODUCTS

As computer technology became available, some data-base developers constructed customized data bases, including many constituents not then present in data bases from government sources or published tables [4]. Some of these large data bases have since been made available to other users with access to mainframe computers.

More recently, the availability of microcomputers has provided data-processing access for most professionals and a segment of the lay public. The number of products available to meet the needs of the second tier of users has expanded rapidly during the past few years. A total of 69 analysis systems are described in the fourth edition of the *Nutrient Data Bank Directory* [7]. Many of the packages were developed for use on microcomputers. The number of foods and nutrients varies. Because initial data storage was limited on microcomputers, some of the early nutrient data bases developed for small machines contain fewer foods and nutrients than are now available. Some developers have concentrated on providing data for popular foods with complete profiles for nutrients of greatest interest.

With so many packages readily available, many individuals are confused when making a choice and do not know how to assess the products. Those users having their first experience with computers are often unaware of what issues to consider. Hence, the first tier of users must assume the responsibility for supplying creditable nutrient data bases and analysis software.

REFERENCES

1. C. F. Adams and L. J. Fincher, *Procedures for Calculating Nutritive Values of Home-prepared Foods: As Used in Agriculture Handbook No. 8, Composition of Foods – Raw, Processed, Prepared, Revised 1963*, ARS 61–13 (USDA, Washington, D.C., 1966).
2. J. T. Andrews, "Development of a Standardized Recipe Data File for Computer Systems," in A. N. Moore and B. H. Tuthill, eds., *Computer Assisted Food Management Systems* (University of Missouri Technical Education Services, Columbia, Mo., 1971), pp. 33–45.
3. J. L. Balintfy, "Menu Planning by Computer," *Communications of the American Computer Machinery*, 7(4): 255–259 (1964).

4. A. A. Hertzler and L. W. Hoover, "Review of Nutrient Databases: Development of Food Tables and Use with Computers," *J. Am. Diet. Assoc.*, 70: 20–31 (1977).
5. L. W. Hoover, "Computers in Dietetics: State-of-the-Art, 1976," *J. Am. Diet. Assoc.*, 68: 39–42 (1976).
6. L. W. Hoover, *Computers in Nutrition, Dietetics and Foodservice Management: A Bibliography* (University of Missouri–Columbia Printing Services, Columbia, Mo., 1983).
7. L. W. Hoover, ed., *Nutrient Data Bank Directory*, 4th ed. (University of Missouri–Columbia Printing Services, Columbia, Mo., 1984).
8. L. W. Hoover and B. P. Perloff, *Model for Review of Nutrient Database System Capabilities*, 2nd ed. (University of Missouri–Columbia Printing Services, Columbia, Mo., 1984).
9. A. Marsh, "Problems Associated with Recipe Analysis," in R. Tobleman, ed., *Proceedings of the Eighth National Nutrient Data Bank Conference* (US Department of Commerce National Technical Information Service, Washington, D.C., 1983), pp. 29–38.
10. R. H. Matthews and Y. J. Garrison, *Food Yields Summarized by Different Stages of Preparation*, Agriculture Handbook No. 102 (USDA, Washington, D.C., 1975).
11. *Provisional Table on Percent Retention of Nutrients in Food Preparation* (USDA Nutrient Research Group, Washington, D.C., 1982).
12. J. Youngwirth, "The Evolution of Computers in Dietetics: A Review," *J. Am. Diet. Assoc.*, 82: 62–67 (1983).

11
Managing Food Composition Data at the National Level

FRANK N. HEPBURN

Nutrition Monitoring Division, Human Nutrition Information Service,
US Department of Agriculture, Washington, D.C., USA

INTRODUCTION

The management of food composition data at the national level is carried out with
the US Department of Agriculture's National Nutrient Data Bank (NDB). A distinction should be made between this management and nutrient data-base management
in the more usual sense, such as is carried out in support of the many computerized
dietary analysis systems – often called nutrient data-base systems – now in operation.
They differ in that the NDB summarizes individual analytical values into a nutrient
data base of representative values for foods. These in turn can serve as the foundation
for the dietary analysis systems. Essentially, the NDB is the provider of summarized
data, and the managers of data systems built upon those summarized data are the
NDB's primary users. It is the purpose of this paper to provide insight into the NDB's
present mode of operation, describe modifications for improvement now under way,
discuss efforts for improving the quality of data, and indicate new applications of
the system that may benefit INFOODS.

The Nutrient Data Bank was conceived and established as a computerized means
of storing and compiling data on the nutrient composition of foods and of providing
average, or representative, nutrient values to data users. Because the computerized
system serves as the mechanism for the revision of *Agriculture Handbook No. 8*,
the expansion of data stored in the NDB parallels progress on the handbook revision.
The current publication status is shown in table 1 [9]. Food groups covered by AH-8,
section nos. 18–22, are those most actively pursued in the data-entering stage at the
present time.

The essential features of the NDB system have been described in detail elsewhere
[5,8]. For this discussion it may be helpful to describe briefly the NDB at each of
its three levels. Data Base 1 (DB1) consists of the individual entries of nutrients
in a food item, together with detailed descriptions of the food item and particulars
concerning the measured value. At present, over 800,000 individual records are stored
in the NDB, and additions continue to be made at the rate of about 6,000 to 9,000
per month.

Table 1. Status of *Agriculture Handbook No. 8* revisions

Sections published [9]	Sections in preparation
8–1 Dairy and egg products	8–13 Beef products
8–2 Spices and herbs	8–14 Beverages
8–3 Baby foods	8–15 Fish and shellfish
8–4 Fats and oils	8–16 Legumes
8–5 Poultry products	8–17 Lamb, veal, and game
8–6 Soups, sauces, and gravies	8–18 Bakery products
8–7 Sausages and luncheon meats	8–19 Sugars and sweets
8–8 Breakfast cereals	8–20 Cereal grains, flours, and pasta
8–9 Fruits and fruit juices	8–21 Fast foods
8–10 Pork and pork products	8–22 Mixed dishes
8–11 Vegetables and vegetable products	8–23 Miscellaneous foods
8–12 Nut and seed products	

Data Base 2 (DB2) consists of summarized values of nutrients in food items that have like descriptions. Individual values are averaged and standard deviations calculated for each grouping. Data at this stage of summary provide the opportunity to examine specific food descriptions, such as year of harvest or region of growth. The application of DB2 information to development of an international data base of cereal grain foods was described in a previous publication [6]. At present, data in DB2 are generally too limited for meaningful statistical distinctions, but the potential for such use by INFOODS should be kept in mind as a means of providing more detailed access to data than is now possible.

Data Base 3 (DB3) contains data at the level familiarly known in *Agriculture Handbook No. 8*. The aim is to provide data that are representative of foods across the nation on a year-round basis. To this end, groupings within DB2 that are indistinguishable at point of purchase or consumption, or that have nearly identical nutrient profiles, may be combined to yield overall mean values. The total number of observations and standard error are also calculated. A provision of the NDB system allows the components to be weighted to produce averages that are more representative for the nation. DB3 is also used to create the computerized version, the USDA Nutrient Data Base for Standard Reference (available from National Technical Information Service, 5285 Port Royal Road, Springfield, VA 22161).

The Nutrient Data Bank is still in its formative period and has not yet reached the stage of continuous maintenance management. At this time, attention is still focused on completing the revision of *Agriculture Handbook No. 8*, and data management is thus devoted primarily to control of data input and output.

DATA INPUT

Data are extracted from the scientific literature on a continual basis, sought from industry, and obtained from government laboratories. In recent years especially, USDA's Human Nutrition Information Service (HNIS) has supported extra-mural contracts for the generation of data on specific foods in order to supply information otherwise lacking. The proportions of data from these sources vary with both the nutrient and the food. The data extracted from scientific literature are usually limited in their range of either nutrients or foods. Reports seldom supply comprehensive data on food composition, and details of analytical methodology and quality control are often incompletely described. Industry has been particularly helpful in providing analytical data upon which label claims are based, but these data are generally limited to the nutrients required for the label and are developed primarily for processed foods. USDA's Nutrient Composition Laboratory (NCL) provides data on selected nutrients in core foods. As discussed by Beecher elsewhere in chapter 18 of this volume, the goals of NCL coincide with ours and we are seeking ways in which to further co-ordinate our activities. The Food and Drug Administration's Revised Total Diet Study [7] is also providing additional, well-documented analyses of foods from known geographic locations.

For work performed under contract we are able to stipulate how samples are to be drawn and handled, the methods to be used for analysis, and precautions to be taken in proper performance. Contractors are required to validate analytical procedures and are asked to develop suitable quality control using standard reference and control materials. Many contractors have voluntarily taken part in analysing control samples routinely examined by co-operators in the National Food Processors Association programme, and for several years have participated in meetings in which they can discuss specific analytical problems and share information on possible solutions. There is no doubt that the co-operation and dedication of our contractors has increased the reliability of results. We realize, however, that absolute control over analytical measurements cannot be attained without imposing quality-control tests supervised by an outside laboratory. A current co-operative project with the Nutrient Composition Laboratory is providing such control in a current study and will serve as a model for future applications.

All data are carefully screened before insertion into the NDB. Data are excluded only with proper justification. The samples analysed must be representative of the food supply, and there must be evidence that the samples have been treated appropriately to avoid contamination or loss of nutrients. The analytical method must be known to be acceptable in the particular application or proven by the researcher. Data that are questionable because of insufficient explanation may be flagged for further evaluation. Flagged data are not included in the computation of means.

The Nutrient Data Bank system is now undergoing major revision. The main purpose is to make the system more efficient, taking advantage of advances in computer technology that have been made since the original system was designed. Two of the new features are of special interest. First, each food-group specialist assigned to work on the NDB will have direct access to the nutrient data through an interactive terminal and will be able to test the effects of different groupings of descriptors in the steps of creating both DB2 and DB3. Second, provision has been made to allow for the attachment of codes to the individual data that will express their reliability in various terms such as adequacy of sampling, methodology, and laboratory quality control. This approach was utilized by Exler when developing a table on the iron content of foods [4]. A similar treatment of selenium data was addressed by Beecher (see chap. 18). It is our expectation that attached codes will be able to be used to develop computer-generated confidence codes for the calculated means.

The range of nutrients included in the NDB follows the interests of the nutrition/health community. Originally it was planned that nutrients should be limited to those for which Recommended Dietary Allowances (RDAs) have been established [3].

Recent observations on possible relationships between dietary components and health have led to demands for data on additional components. The report on *Diet, Nutrition and Cancer* [2], for example, pointed to the possible roles of carotene, dietary fibre, and selenium. Because of the lag between expressed interest in a new food component and the ability to generate data for its content in foods, we must anticipate users' needs and take an early initiative to supply new information. In our search for data, we concentrate on those components currently covered in the revised handbook, plus those of growing importance. Data for additional components, although not actively pursued, are entered when included in analytical reports.

The question of which foods should be included in the data bank is similar to the question about nutrients, and is perhaps as complicated. To serve as a national source of information we must be sure to cover those foods most frequently consumed by the general population, but we cannot neglect others that may be important only to certain population subgroups. Foods reported in the Nationwide Food Consumption Surveys provide both types of information. We try to keep abreast of trends in food production, the introduction of new processed foods, changes in formulation or processing, the introduction of new cultivars, and changes in breeding or feeding practices, in order to anticipate possible changes in our nation's food supply and to prepare for the impact such changes may have on measuring nutritional components.

DATA OUTPUT

The primary product of the NDB is the revision of *Agriculture Handbook No. 8*. In machine-readable form this is represented by USDA Nutrient Data Base for Standard Reference, which basically consists of the data contained in the revised sections of the handbook, supplemented by the older data for food groups that has not yet been revised. The Standard Reference tape is updated as new revisions are released and thus always represents the most up-to-date information available. Each release of the USDA Nutrient Data Base for Standard Reference is identified by a release number and the year. The most recent is Release No. 5, 1985, which covers revised sections of the handbook through no. 8–12 [9].

The USDA Nutrient Data Base for Standard Reference is utilized for the creation of specialized data bases, both within USDA and by users who purchase the tape for adaptation to their specific needs. A good example of a data base derived for USDA use is the Nutrient Data Base for Individual Food Intake Surveys, which is in the final stages of completion. This has been created by developing a computerized linking file that connects the survey food codes to the food codes on the USDA Nutrient Data Base for Standard Reference. For survey items that are composite foods, formulas are included by which values for the composite items can be calculated from the nutrient content of the individual components. For the forthcoming Continuing Survey of Food Intake by Individuals (CSFII), we were asked to include data for components not regularly present in the Standard Reference base. This made it necessary to expand it to include data for dietary fibre, alcohol, vitamin E, and carotenes for approximately 1,700 food items. In addition, 4,000 values for other nutrients were added to supply values not yet contained in the Standard Reference data base because the food groups have not yet been updated. We have named this expanded data base the Primary Data Set for Food Consumption Surveys. The linking file, the Primary Data Set,

Table 2. Proportion of nutrient values based on analytical data in Primary Data Set

Nutrient	Percentage of values
Protein	98
Calcium	95
Magnesium	78
Carotene	54
Dietary fibre	29

and a computerized table of retention factors are accessed by a computer program to create the Nutrient Data Base for Individual Food Intake Surveys.

In creating the Primary Data Set we were placed in the unique position of being both data users and data providers at the same time. Care has been taken to document the sources of the expanded data so that they can be further evaluated and updated as additional information becomes available. Codes have been attached to all added nutrient values to indicate whether they are from analytical data in the revised handbook, new analytical data not yet finalized for the handbook, older data from sections of the handbook not yet revised, or whether they are imputed values or assumed values of zero, such as for cholesterol in plant foods.

Besides documenting the data sources, this coding system provides a new way to measure the state of knowledge of food composition data. A quantitative measure of available analytical data for each nutrient under consideration can be calculated by determining the relative proportion of analytical to imputed values in the expanded data base. Examples of such calculations are shown in table 2. Calcium and protein are representative of nutrients that have been analysed regularly over a long period and for which analytical data are thus most available. At the other extreme is dietary fibre, for which analysed values are just beginning to be reported. Although analyses for magnesium are now commonly included in food composition studies, only limited information was available for the 1963 edition of *Agriculture Handbook No. 8*, and it is apparent that analytical values currently available are not as comprehensive as those for calcium. Analysed values on hand for carotene are almost entirely those determined in plant products by AOAC procedures [1]. For the Primary Data Set the remaining values for carotene were those assumed to be present in arriving at expressions for total retinol equivalents of vitamin A.

It must be understood that the calculations shown in table 2 pertain directly only to the Primary Data Set. They indicate the basis for information on hand for calculating the composition of foods in the CSFII data base. It should be further understood that the procedure distinguishes only between analytical and imputed data, without attempting to address the reliability of the analytical data.

SPECIAL CONSIDERATIONS

Determining when a value in a data base should be changed is a major problem in managing the NDB. The problem is universal and should be addressed by INFOODS to insure common approaches to the solution.

There are two major aspects to the problem. The first concerns data reliability.

There is general recognition that all data are not created equal, because of either deficiencies in the analytical methods themselves or inappropriate application of methods. The concerted efforts being directed toward improving data reliability will gradually make available better values, but for some period the new data will coexist with those now on hand. We must learn how to deal with changes warranted by the availability of new data, and in the interim it is more important than ever to document the sources of data, recognizing whatever shortcomings they may have.

A second aspect of recognizing change is caused by actual changes in the food supply. The introduction of new cultivars, adoption of new feeding practices, and technological changes in food processing are all capable of altering the composition of food. Two examples are the recent development of more highly coloured yellow vegetables, with a resultant increase in carotene content, and the reduction in sodium in some processed foods. Not only must such changes be recognized, but the time that the change occurred must somehow be accounted for. This is of particular importance to HNIS in proceeding with its continuing survey, and it will also be important in comparing changes in the food supply available in different regions of the world.

CONCLUSIONS

The management of food composition data in the United States should be no different than in any other country and should be representative of the situation confronting INFOODS. Certainly the successful solution to problems depends upon the co-operative interaction between data users and data providers.

REFERENCES

1. Association of Official Analytical Chemists, *Official Methods of Analysis*, 14th ed. (Association of Official Analytical Chemists, Inc., Arlington, Va., 1984).
2. Committee on Diet, Nutrition, and Cancer, Commission on Life Sciences, National Research Council, *Diet, Nutrition, and Cancer: Directions for Research* (National Academy Press, Washington, D.C., 1983).
3. Committee on Dietary Allowances, Food and Nutrition Board, Commission on Life Sciences, National Research Council, *Recommended Dietary Allowances*, 9th ed. (National Academy Press, Washington, D.C., 1980).
4. J. Exler, *Iron Content of Food*, Home Economics Research Report, no. 45 (Human Nutrition Information Service, USDA, Washington, D.C., 1982).
5. F. N. Hepburn, "The USDA National Nutrient Data Bank," *A.J.C.N.*, 35: 1297–1301 (1982).
6. F. N. Hepburn and B. P. Perloff, "The Nutrient Data Bank," *Cereal Foods World*, 24: 224–225 (1979).
7. J. A. T. Pennington, "Revision of the Total Diet Study Food List and Diets," *J. Am. Diet. Assoc.*, 82: 166–173 (1983).
8. R. L. Rizek, B. P. Perloff, and L. P. Posati, "USDA's Nutrient Data Bank," *Food Tech. in Australia*, 33: 112–114 (1981).
9. US Department of Agriculture, "Composition of Foods: Raw, Processed, Prepared," *Agriculture Handbook No. 8* (Science and Education Administration, USDA, Washington, D.C., 1976–1984).

12
Maintaining a Food Composition Data Base for Multiple Research Studies: The NCC Food Table

I. MARILYN BUZZARD and DIANE FESKANICH

Nutrition Co-ordinating Center, Division of Biometry, School of Public Health, University of Minnesota, Minneapolis, Minnesota, USA

INTRODUCTION

A food composition data base was developed at the Nutrition Co-ordination Center (NCC), University of Minnesota, in 1974 for use in analysing dietary data for two long-term multi-centred cardiovascular studies, the Multiple Risk Factor Intervention Trial (MRFIT) and the Lipid Research Clinics (LRC) Programs [2–4]. Standardized methods for the collection and analysis of dietary data and centralized processing of the data were implemented to minimize inter-clinic differences among the 34 centres involved and to allow comparability of dietary data between the two collaborative research programmes supported by the National Heart, Lung and Blood Institute (NHLBI).

The original nutrient data base and coding system were designed to allow detailed specification of both the quality and quantity of dietary fat. Margarines, oils, and shortenings were classified by type and brand, and fats used in recipes and food preparation were documented in detail, as described by Dennis et al. [1]. Within food groups, food items with similar fat content were grouped together. The original data base included approximately 1,200 entries with values for 31 nutrients.

In 1977 the NCC nutrient data base coding system became available for other research studies, and the system has been used by numerous investigators over the past eight years. The majority of users are medical researchers in the United States and Canada involved in the investigation of relationships between diet and disease. The data base has been expanded to meet the research needs of each study. The current NCC nutrient data base (referred to here as the NCC Food Table) includes approximately 1,800 entries and values for 61 food components.

The purpose of this paper is to describe the needs of the users of the NCC Food Table and to discuss the NCC approaches to maintaining the table to meet these needs. NCC guidelines for minimizing redundancy, without loss of the specificity required by the users of the system, are also described.

SPECIFIC USER NEEDS AND APPROACHES TO THESE NEEDS

The Need for Standardized Methods of Determining Nutrient Values for All Foods and Beverages Consumed by the Study Population

Users of a nutrient data base require that the system be capable of handling not only common foods and beverages, but also any uncommon items that might be consumed by the study population. New entries are frequently added to the NCC Food Table to meet study-specific needs. Food composition data are gathered for regional foods and local recipes which appear on dietary intake forms, and also for new manufactured products or reformulations of existing products consumed by the study population. These new foods or beverages are compared with existing Food Table entries. If the item cannot be represented by an existing entry or a combination of existing entries, a new entry is developed. All coding decisions are documented and cross-referenced as appropriate to ensure that the item will be coded in the same way if it appears on future dietary intake forms.

The Need for Updated and Complete Nutrient Profiles

Nutrient data bases must be updated continually to meet the needs of a wide range of investigators. The updated file must reflect new published or provisional data provided by the United States Department of Agriculture (USDA), new analytic data from the scientific literature, and current food composition data for new or reformulated commercial products.

To ensure that nutrient values in the NCC Food Table are current, several dozen scientific journals are reviewed by the NCC nutrition staff on a monthly basis. Any new data on food composition, new products, or food labelling are extracted. These data, along with any other newly available data from USDA or food manufacturers, are compared with existing values in the Food Table. Limits have been established for each nutrient to guide data-base nutritionists in determining when the difference between the new and the existing value is sufficient to warrant a modification to the Food Table; modification procedures are implemented only when differences exceed the established limit values. This system prevents the expenditure of considerable time and effort on making changes to the Food Table that are insignificant considering the variability inherent in nutrient data.

Major food manufacturers are contacted annually for updated information on the nutrient composition of their products, and new data are compared with existing values. Differences that exceed the established limit values are identified as being caused by either product reformulation or better analytical data. If the product has been reformulated, a new entry is added to the Food Table and the old entry deactivated so that it cannot be used in future coding. If the nutrient differences are the result of better analytical data, modification procedures are implemented.

New nutrients have been added to the NCC Food Table to accommodate investigators' current research interests. USDA data are the preferred sources, but when these are incomplete or unavailable other sources, including foreign food tables, scientific literature, data from food manufacturers, or unpublished laboratory data, are used. All sources of raw data are reviewed and compared by a data-base nutritionist, and values are selected or imputed for the Food Table based on factors which include the reliability of the data sources, the laboratory analytic method used, and the number of samples analysed.

Adding new nutrients to the NCC Food Table often requires making separate entries for items grouped within a single entry. For example, the addition of sodium to the nutrient profile resulted in new entries for canned vegetables, which were previously included in the entries for vegetables cooked from fresh or frozen sources.

Users also need a data base that is complete for all nutrient values. Since missing values are calculated as zeros in the nutrient analysis, they are tolerated only in cases where zero is as good an estimate as any other for a given nutrient in a given food item. The NCC data-base nutritionists impute values, whenever possible, from the nutrient content of similar foods, recipes, or product ingredient lists. As soon as better data become available, the values are updated to reflect the analytic data.

The Need for Documentation of Sources of Nutrient Values

Users of a nutrient data base require documentation of the sources of nutrient information used in the calculation of their dietary data to enable them to judge the reliability of the nutrient analysis.

The NCC reference system includes a six-character field for proximate composition and a two-character field for all other food components. The leading character in the proximate composition field designates the general category of the reference source, and remaining characters specify the actual page or code number in the designated source. The two-character code for nutrient references designates the source as an USDA reference, other food table reference, specific journal reference, or other source of nutrient data, including calculations and imputations.

A paper file is maintained for every entry in the NCC Food Table. All details of calculations and imputations of nutrient values, including the rationale for selection of the foods on which these determinations are based, are documented in the paper file. General guidelines by food groups are maintained for calculating or imputing nutrient values.

Codes for reliability of nutrient data have been discontinued due to the difficulty of establishing objective guidelines to meet the needs of all users. Sources of all nutrient data are documented as completely as possible in the NCC Food Table so that the user is able to judge reliability in relation to a specific research setting.

The Need for Quality-control Procedures to Ensure the Accuracy of the Data Base

Nutrient data-base users need assurances that the values used in the calculation of their dietary intake data are accurate and that the computer programs used to calculate the data are reliable.

Despite ongoing efforts to maintain the accuracy of a nutrient data base, mistakes can find their way into a system as a result of human error or mechanical or software malfunctions. A number of quality-control procedures have been developed by the NCC to document the accuracy of the Food Table and the calculation software.

The NCC Food Table is updated on a daily basis by the data-base nutrition staff. All Food Table modifications are made to a separate file called the Reference Food Table. Nutrient modifications are edited in the data-entry system by comparing new nutrient values with NCC established nutrient ranges. A value that falls out of range must be verified by an NCC nutritionist.

The Reference Food Table is periodically checked for accuracy by running a series of computer-generated integrity reports that are reviewed by the nutrition staff. Some integrity checks are algorithms that are calculated for each entry and compared with a predetermined value. Other integrity checks are listings of individual nutrients by food groups. Any nutrient value that deviates substantially from other values in that food group is verified by a nutritionist.

After all integrity reports have been verified, a test set of dietary intake records are analysed to check the various calculation procedures. The results of the nutrient calculations are compared with the previous calculations of the test records. Any differences observed must be verified as being due to recent modifications to the Reference Food Table. After satisfactory implementation of these procedures, the Reference Food Table is available for research use and becomes the current version of the NCC Food Table.

The Need for Stability of the Nutrient Data Base for Long-term Studies

Although the majority of researchers want to use the most current nutrient information available at the outset of a study, they require that the data base remain stable for the duration of their study. This may present a problem for those who maintain a nutrient data base for multiple users.

The NCC has resolved this problem by maintaining multiple versions of the Food Table. The most recent version available at the beginning of a long-term study is used for nutrient calculations for the duration of the study. The only permissible change to a study-assigned Food Table is the addition of new entries for foods or beverages not previously included. It is essential that no modifications be made to previously established entries; changing the data base while a study is in progress will confound the interpretation of the dietary data. At the end of the study, the investigator may choose to rerun all data on an updated version of the Food Table.

The Need for Comparability of Nutrient Data between Studies

Investigators may require a nutrient data base that will permit comparison of their study results with other studies using the same data base. Inter-study comparability of dietary data can provide information beyond the scope of a single study.

A system has been established at the NCC that allows comparison of dietary data between past and future studies. Such comparison is possible even though the NCC Food Table is updated routinely. This is accomplished by ensuring that no entry is ever deleted from the data base. When a product is taken off the market, the item is "deactivated," which means that it can no longer be used in coding; however, the item remains in the data base and its nutrients continue to be updated as new nutrients are added to the Food Table. Maintaining these deactivated entries in the Food Table makes possible the recalculation of nutrients for dietary intake data collected in the past while using an updated version of the Food Table. The results of these recalculations can then be compared with those of current studies.

The Need for Flexibility in the Level of Specificity
Required for Documenting Dietary Detail

Some researchers require considerable detail in documenting dietary intake data while others select methodologies that document dietary intake in broad food group categories. A nutrient data base that meets the needs of diverse users must be able to provide nutrient calculations for dietary intake data documented at different levels of specificity.

The NCC has met this need for flexibility by providing procedures that capture the highest level of detail (such as the brand names, preparation methods, and salt additions) or revert to default values when the higher levels of detail are not specified. Nutrient values for the default or "unknown" assignments are based either on weighted averages of values for the items that fall into that general category or on a single item that is representative of the items in that category.

Guidelines for creating defaults using weighted nutrient values or selected representative values are based on food intake patterns of specific study populations. Determination of defaults for fats used in food preparation methods takes into consideration whether the item was commercially processed, prepared at home, or prepared in a restaurant; if the latter, the default assignments are based on the price range of the restaurant.

The Need for Flexibility in Specifying Food Quantities for Data Input

Researchers need to be able to report dietary data as described by study participants without having to transform food quantities into specified units. The NCC Food Table meets this need by maintaining densities and/or weights of food-specific units. For entries that contain density data, food quantities can be reported using any cubic or household measure of volume. The nutrient calculation programmes convert volume measurements to a gram weight using the food density. For foods that are not easily described in common volume measurements, such as stalks of celery, slices of bread, or pieces of pie, weights are provided for food-specific servings. Dimensions are described for each food-specific serving.

MINIMIZING REDUNDANCY IN THE NUTRIENT DATA BASE

Ongoing maintenance and expansion of a nutrient data base to include increasing numbers of nutrients and other food components can be efficiently accomplished only if the number of entries requiring routine maintenance can be kept to a minimum without losing the specificity required by the many users of the system.

All entries in the NCC Food Table are either elemental or composite entries. Elemental entries may be defined as entries for which nutrient values are maintained in the Food Table. Composite entries are defined by an ingredient list of two or more elemental entries with specified amounts. Nutrient values for composite entries are calculated from the ingredient list by the computer. This system limits the routine maintenance of the Food Table to the elemental entries.

The following guidelines have been developed by the NCC to minimize redundancy in the Food Table by limiting the number of elemental entries.

1. Include Foods Only in the Forms in Which They Are Eaten

Since analysis of the dietary intake of individuals living in the United States and Canada is the common need of all current users of the NCC system, the Food Table need not include foods in forms in which they are never consumed by the study populations. Thus, no entries are included for most raw meats, uncooked pasta, or certain raw vegetables such as potatoes. Many foods commonly eaten in other countries do not appear in the NCC Food Table because they have not been encountered frequently enough on the dietary records received at the NCC. As various ethnic foods increase in popularity in this country, it is expected that more foreign foods will be added to the Food Table. Sushi is an example of a recent addition to the NCC Food Table.

For composite entries of cooked recipes, cooked ingredients are substituted for the raw ingredients whenever possible. For example, if a casserole is made with raw rice, the corresponding amount of cooked rice is entered as a recipe ingredient in the composite entry. Use of the nutrient values of cooked rather than raw ingredients makes the calculated nutrient content more similar to the actual nutrient content of the food as eaten. This system also reduces the number of elemental entries required in the Food Table by eliminating the need to maintain raw food entries as recipe ingredients for composite entries.

2. Combine Foods of Similar Nutrient Content into a Single Entry

When similar foods differ by less than the established limit values for each nutrient, the foods are grouped together in a single entry. For example, spinach cooked from fresh and cooked from frozen are combined in a single entry. If at some point a new nutrient or other food component is added to the Food Table for which the value in cooked fresh spinach differs from that in cooked frozen spinach by more than its established limit value, the two items would be given separate entries.

3. Add New Foods as Composite Entries Rather than as Elemental Entries Whenever Possible

Even though some mixed dishes have been analysed for nutrients in the composite state, it is preferable to add them to the Food Table as composite entries rather than as elemental entries as long as the amounts of ingredients can be specified and the calculated nutrient values closely match the values provided for the composite item. Most home-prepared foods in the NCC Food Table are maintained as composite entries. Some commercial products with well-defined ingredients are also entered into the Food Table as composite entries. Approximately one-third of the current NCC Food Table consists of composite entries.

4. Use of Prep Codes and Fat Codes

"Prep codes" to specify amounts of fat added in various food preparation methods and "fat codes" to designate the type of fat used are other procedures used by the NCC to limit the number of elemental entries. These procedures have been described in detail elsewhere [1]. Prep codes prompt the appropriate computer algorithms to calculate the amount of fat, salt, or other additions for various food preparation methods of a basic food. Thus a single entry can be used for many different preparations of that item. For example, a piece of light-meat chicken without skin can be breaded and fried in corn oil, baked with butter, or broiled without added fat by invoking the appropriate prep codes and fat codes with the same elemental item.

5. Use of Add-Principal-Fat (APF) Recipes

Foods that contain significant ingredient or cooking fats are designated as APF recipes and are maintained in the NCC Food Table as composite entries. French fried potatoes, pie crust, and salad dressing are examples. All APF recipes require specification of the type or brand of the predominant fat ingredient. Thus, the nutrients in corn bread could be calculated using bacon fat, soybean oil, shortening, or any other appropriate fat. This system allows considerable flexibility for specifying ingredient and cooking fats without increasing the number of entries in the Food Table.

6. Use of Coding Guides for Brand-name Products and Food Characteristics

Coding guides are alphabetical indices of specific types, classes, or brands of foods that designate the particular NCC Food Table entries into which they are classified. For example, the Brand Name Margarine Guide specifies which of the approximately 60 margarine entries in the Food Table should be used for each of approximately 400 brand-name margarines. The Beef Guide specifies which Food Table entry should be used for each type or cut of beef. The current NCC system includes approximately 80 guides. Each guide also includes directions for coding items when the type or brand is not specified.

7. Handling of "Uncodables"

When an item appearing on a dietary intake record cannot be coded according to established procedures, the item is documented on an Uncodable Form to be resolved by the nutrition staff. If the uncodable item is a new product on the market, ingredient and nutrient information is requested from the manufacturer to determine proper coding. If the uncodable item is a composite food consisting of unknown amounts of ingredients, the nutritionist makes a judgement on proportions. Decisions on the coding of uncodable items are stored in a cross-referenced file to facilitate standardization for future coding decisions. Uncodables that begin to appear frequently on intake records are added to the Food Table as new entries. An example of an uncodable recently converted to a new entry is trail mix. The uncodables system prevents the unwieldy expansion of the Food Table that would result from the inclusion of many infrequently consumed items.

Use of the various procedures outlined above effectively limits the number of elemental entries in the NCC Food Table. Thus, updating can be routinely implemented and new nutrients added with minimum effort while maintaining maximum flexibility for detailed specificity to meet user needs.

SUMMARY

A table of food composition designed to meet the ongoing nutrient analysis needs of multiple research studies must be continually updated and expanded. Standardized methods of updating and imputing nutrient values must be established, and sources of all values must be carefully documented. Computerized edit checks and other quality-control procedures must be incorporated into the system to ensure the accuracy of the data base.

To accommodate the needs of multiple long-term studies, a number of versions of the data base are maintained by the NCC. Flexibility is provided to meet different levels of specificity of dietary detail required by different research protocols.

Procedures have been established by the NCC to facilitate ongoing maintenance of the data base without loss of the specificity required by the users of the system.

REFERENCES

1. B. Dennis, N. Ernst, M. Hjortland, J. Tillotson, and V. Grambsch, "The NHLBI Nutrition Data System," *J. Am. Diet. Assoc.*, 77: 641–647 (1980).
2. Lipid Research Clinics Epidemiology Committee, "Plasma Lipid Distributions in Selected North American Populations. The LRC Program Prevalence Study," *Circulation*, 60: 427 (1979).
3. Lipid Research Clinics Program, "The Coronary Primary Prevention Trial. Design and Implementation," *J. Chronic Dis.*, 32: 609 (1979).
4. Multiple Risk Factor Intervention Trial Group, "Statistical Design Considerations in the NHLBI Multiple Risk Factor Intervention Trial (MRFIT)," *J. Chronic Dis.*, 30: 261 (1977).

13
Managing a Nutrient Data-base System: Meeting Users' Needs and Expectations

GRACE J. PETOT

Departments of Nutrition, Epidemiology and Biostatistics, School of Medicine, Case Western Reserve University, Cleveland, Ohio, USA

INTRODUCTION

Approximately 25 years ago, the first computerized food composition tables were used for dietary evaluation in epidemiological studies [8, 21]. Computer processing of dietary data demonstrated improved comparability of data from different sources, easier application of new approaches to analysis and more efficient storage of data for future use [9]. Tremendous advances in computer hardware and software development, along with the burgeoning interest in food, diet, and health, have expanded the applications for food composition information far beyond those of 25 years ago.

The purpose of this paper is to describe a large nutrient data-base system and to document those activities involved in managing it to meet users' needs and expectations.

THE HVH-CWRU NUTRIENT DATA-BASE SYSTEM

The Highland View Hospital–Case Western Reserve University (HVH-CWRU) Nutrient Data Base was developed in the early 1960s to study the diets of chronically ill people who were living at home. At that time, a suitable nutrient data base was not available and, because of the large number of individuals to be studied over time, it was decided to create a food composition table using the computer [11]. The multi-disciplinary development team included a physician, a research dietitian, a biostatistician, and a systems analyst, each of whose interest was in the use of the data generated as well as the generation and processing of the data. The team approach to the continuing development and management of the system has remained until the present time. By the mid-1960s, it was recognized that there were other research efforts that would require more extensive food composition information; however, sources of funds were not found. In spite of this, the group decided to completely revise the original data base, with local support for development and research from the Division of Nutrition at Highland View Hospital and the Department of Biometry at Case Western Reserve University. Since 1978, with the dissolution of the Division of Nutrition at Highland

View Hospital, the data base has been maintained by the Departments of Biometry (at present the Department of Epidemiology and Biostatistics) and Nutrition in the School of Medicine, Case Western Reserve University.

The primary source of food composition data was, and remains, the United States Department of Agriculture (USDA) *Handbook No. 8*. Machine-readable forms of the data were not available during the early development years, and the data were therefore transferred by hand to computer cards. Since USDA data were not always adequate to meet the various research needs, other sources of information were used. Criteria for inclusion of data were established, the most important of which was that the data base reflect current knowledge of the nutrient composition of food. These criteria are, of course, fundamental to the validity of the entire process. For nutrients or for foods not included in *Handbook No. 8*, information is selected from research reports in refereed journals, directly from the food industry for brand-name foods, and from calculations of recipes based on nutrient values for ingredients. Close communication is maintained with the Food Composition Group at USDA for advice and consultation regarding the reliability of all data.

The food composition table at present contains more than 3,000 food items and recipes. For each food item, there is storage space for the nutrient values per 100 grams of edible food product, shown in table 1. Information stored with each food item is illustrated in table 2. The food items are placed in one of 44 food groups and are arranged alphabetically into subgroups related to common attributes of form, preparation, processing, or nutrient content. The coding manual for the NVH-CWRU

Table 1. Nutrients included in the HVH-CWRU nutrient data base

Calories	Riboflavin	*Amino acids*
Total protein	Niacin	Cysteine
Animal protein	Pyridoxal B$_6$	Cystine
Plant protein	Vitamin B$_{12}$	Histidine
Total fat	Folic acid	Isoleucine
Animal fat	Pantothenic acid	Leucine
Plant fat	Biotin	Methionine
Total carbohydrate	Choline	Phenylalanine
Refined carbohydrate		Threonine
Natural carbohydrate	*Minerals*	Tryptophan
Alcohol	Iron	Tryosine
Ash	Calcium	Valine
Fibre	Phosphorus	
Water	Sodium	*Fatty acids*
Caffeine	Potassium	Total saturated
Cholesterol	Magnesium	Total unsaturated
	Chlorine	Oleic
Vitamins	Chromium	Linoleic
Vitamin D	Cobalt	
Total vitamin A	Copper	*Sugars*
Preformed vitamin A	Iodine	Glucose
Beta-carotene	Manganese	Fructose
Total tocopherol	Molybdenum	Lactose
Alpha-tocopherol	Selenium	Maltose
Other tocopherol	Sulphur	Sucrose
Ascorbic acid	Zinc	Reducing sugars
Thiamine		

Table 2. Information stored for each food item in the HVH-CWRU nutrient data base

Food group code	2 digits
Food item code	4 digits
Food name with attributes	60 characters
Major source code for nutrient composition data	7 characters
Presence or absence of lactose and gluten	2 characters
Volume or household measure code	2 digits
Volume or household measure weight in grams	Value between 0.0001-99999
Alternate volume or household measure code	2 digits
Alternate volume or household weight in grams	Value between 0.0001-99999
Nutrient value for each of 71 nutrients	Value between 0.0001-99999
Source code for each nutrient value	1 character

Nutrient Data Base contains a listing of all food items with identification codes, volume or household measure codes. For all food items, nutrient values may be retrieved for any one of six or seven household measures. The variety of measure codes provides considerable flexibility in expressing household or volume measures of food items and reduces professional and clerical time in preparation of data for nutrient analysis. To obtain nutrient data for an item, the following information obtained from the coding manual must be provided to the computer: (a) the food item identification number; (b) a permissible measure code; and (c) a quantity or amount of the measure.

Two versions of the nutrient data-base software are maintained. A Fortran 77 version is transportable to a large number of mainframe or mini-computers with Fortran compilers. The C version is for computers running the UNIX operating system and features simplified coded data input, file storage, and manipulation of data output.

Standard system calculations include nutrient summaries for lists of foods, menus, dietary intakes, etc., averages for up to 99 summaries, percentage of recommended dietary allowances (RDAs) for each age, sex, pregnancy or lactation category, percentage distribution of energy for total protein, total fat, total carbohydrate, animal and plant protein, animal and plant fat, refined and natural carbohydrate, polyunsaturated and saturated fatty acids. Electrolytes are expressed in weight units and in milliequivalents.

The nutrient data-base system has been made available to a large number of users for the past ten years; it has been purchased by 15 institutions and one practising dentist. Although close communication is maintained with active owners, it has not been possible to document all of their applications. Major categories of their uses are clinical practice and research, education, market research, food-product development, and nutrient-analysis services. The institutions owning the data-base system are university academic departments, university medical centres, hospitals, food manufacturers, and a food trade association.

USES AND USERS

Eight broad categories of uses of the data system have been identified (table 3). Clinical research utilizing dietary intake represents the area of greatest use for investigations of the relationships between diet and health. These include studies of dietary components and hypertension [22], weight reduction with drugs, anorexia, dialysis in renal

Table 3. Uses and users of the HVH-CWRU nutrient data base

Clinical research	*Professional or academic education*
University academic departments	University academic departments
University medical centres	University medical centres
Food manufacturers	Health promotion organizations
Dialysis clinics	State health departments
Clinical psychologists	Software developers
Hospital clinical research centres	Voluntary health organizations
Long-term care institutions	Food trade associations
National Institutes of Health	Professional associations
Graduate students	
Pharmaceutical manufacturers	*Patient education*
	University academic departments
Epidemiological	Dialysis clinics
	Clinical psychologists
University academic departments	Dentists
University medical centres	Physicians
Research contractors	Health promotion organizations
	Consulting dietitians
Clinical practice	Voluntary health organizations
	Visiting nurses' associations
University medical centres	Software developers
Dialysis clinics	Professional associations
Clinical psychologists	
Dentists	*Information for the public*
	University academic departments
Food-service menus and recipes	Food manufacturers
	Food service contractors
University academic departments	State health departments
Food manufacturers	Voluntary health organizations
Hospitals	Software developers
Food service contractors	Food trade associations
Long-term care institutions	Professional associations
Armed services	Supermarket chains
Correction/penal institutions	Daily newspapers
State health departments	Secondary schools
Software developers	Recipe-book authors
Food brokers/distributors	Libraries
Pre-schools	
	Rule-, policy-making
Food product development	Federal agencies
Food manufacturers	
Physicians	

disease, pulmonary disease, alcohol consumption [3, 4, 14], aging [5, 6, 7], cervical dysplasia [23], multiple sclerosis [12, 26], mental retardation, dental caries [2], atherosclerosis, high-risk pregnancy, and lactation.

The majority of these studies have been conducted at university medical centres, clinical research centres, and academic departments; however, some have been done by pharmaceutical companies investigating the nutritional adequacy of dietary supplements or enteral feeding products and by food manufacturers for in-house research for new product development. A number of clinical research investigations have been completed, or are in progress, by doctoral and masters' candidates in nutrition and medical computer systems.

Epidemiological surveys by independent research contractors with government agencies have been completed or are in progress. The largest survey was the National Evaluation of School Nutrition Programs, contracted by the Food and Nutrition Service, USDA, with System Development Corporation, in which approximately 7,000 24-hour recalls were coded and analysed for 26 nutrients [25].

The primary requirement for clinical and epidemiologic dietary studies is for current, valid, reliable nutrient calculations for specific foods, beverages, or special dietary products, in specific quantities as consumed by persons of all ages living in any region of the United States. Food-intake information is usually received as 24-hour recalls, quantified diet histories, food-intake diaries or observed records of intakes. Portion sizes may be weighed or, more frequently, estimated. There is a notable lack of standardized protocols for obtaining this information, and therefore it represents many levels of completeness. The records may be obtained by trained individuals, but in many cases are self-reported. Therefore, depending on the quality of such records, it becomes the responsibility of the data-base management team to qualify the results in terms of the completeness of the collected information.

Evaluations of the nutritional quality of analysed records, other than comparisons with the RDAs, are not provided with nutrient-analysis services. It has been the policy of this management team to recommend consultation with dietitians or nutritionists about recording of data and for interpretation of results. In spite of this, some investigators continue to plan and conduct dietary studies without recognizing the need for such assistance.

Clinical applications of nutrient analysis have been implemented in a number of large hospitals. University Hospitals of Cleveland, a 1,000-bed teaching institution, accesses the data base daily for assessing the dietary intakes of about 100 patients [16, 17]. The hospital dietary department and the data-base management team have developed a recipe file system for storing the coded ingredients of each recipe. This file provides access to current nutrient values for each recipe without actually storing pre-calculated nutrient values. Efficient procedures, using computer generated pre-coded forms for recording intakes, are used. Completed reports are then delivered to dietitians within hours. As costs of purchasing and maintaining hardware decline, and as software becomes more transportable, it is anticipated that many more institutions will include dietary nutrient analyses as a routine component of nutritional assessment.

A major use of the data-base system is for the nutrient analysis of menus and recipes for institutional food service. These analyses usually serve as documentation of the nutritional adequacy of food served in hospitals, nursing homes, correctional institutions, state-supported residential institutions, schools, and the armed services. Governing or accrediting agencies for these institutions have set nutritional standards and usually require a comparison with the RDAs as evidence of adequacy. Rarely do they request analyses for other nutrients or food constituents. Independent developers of software are incorporating nutrient analysis into computerized systems for hospitals, restaurants, and food-service contractors for management of food inventories, recipes, menus, production schedules, cost accounting, or clinical services. Many developers have realized that creating and maintaining a food composition table is costly and are purchasing food tables with or without programs for accessing and for formating reports.

Another user of a nutrient analysis system is the broker or distributor who offers menu and recipe analyses as a service to purchasers of food products. Food manufacturers have found that using a computer for examining and manipulating ingredient and food-product formulas for optimal nutrient content saves time and money in the development of new products [13]. Calculations of nutrient content per portion of many prepared, ready-to-heat-and-eat food-service products are supplied to institutions with the products purchased. It is not necessary for these analyses to meet the Food and Drug Administration nutrition labelling regulations, which require laboratory analysis if nutrient content appears on the label. Several physicians have formulated special dietary products for use in their practices and are using nutrient calculations during the development and testing phases.

As the computer replaces printed food composition tables, dietetic, nutrition, and food science students have been introduced to a very effective method for learning what is in food. The fast response time for queries and manipulations of menus, recipes, and portion sizes provides more information with greater understanding of the inter-relationships of foods, nutrients, portion sizes, and diets. This capability extends to the potential provision of nutrient information to practising professionals through nationwide access to nutrient data-base systems. Professional and trade associations and health promotion organizations are exploring the feasibility of providing this access to members and clients.

At another, less sophisticated level, nutrient-analysis services are being made available to the public. Health-care professionals are using simplified, understandable nutrient summaries of patients' dietary records or of their favourite foods. Food exchange lists are easily developed to meet a variety of diet prescriptions. The American Heart Association uses nutrient analysis services in the development of educational materials for physicians, dietitians, patients, and the general public [1].

Recently, through an agreement with a local software developer, a nutrient-analysis system designed for a variety of personal computers has been made available at low cost to the home user. It incorporates a data base of 800 foods and recipe ingredients with 15 nutrients which were extracted from the HVH-CWRU Nutrient Data Base. It allows the home user to plan and evaluate the nutritional adequacy of recipes, menus, and dietary intakes. Other purchasers of this software include physicians, dentists, libraries, elementary schools, secondary schools, and universities. An interesting feature of this version is the expression of refined carbohydrate in teaspoons in the nutrient summaries. A professional package for use on microcomputers, with the same food table, is being marketed to small hospitals, nursing homes, consulting dietitians, physicians, and dentists.

Two federal agencies have used the HVH-CWRU Nutrient Data Base to examine the nutrient content of common or usual portions of food [18, 19, 20]. Reports generated included nutrients per portion, percentage US RDA per portion, Index of Nutritional Quality, and nutrients per 100 kilocalories. This information was used by the Federal Trade Commission for proposing rules regulating the advertising of a "nutritious food" and by the Food and Nutrition Service of USDA for regulating the kinds of foods sold in competition with federally subsidized school meals.

MEETING USERS' NEEDS AND EXPECTATIONS

All users of the nutrient data-base system express the need for: (a) reliable and valid food composition data with (b) easy access in as short a time as possible, (c) reported in the desired form, and (d) at little cost. An unexpressed need for many is one of hand-holding, consultation, education, and encouragement.

Food Composition Table and Software

It has already been shown that food composition data are used for many purposes. It then becomes evident that the degree of precision and accuracy of nutrient data reports is dependent upon the accuracy and completeness of the quantitative and descriptive information provided for analysis. Users of data for clinical research and applications have the greatest need for complete and accurate analyses. Experience indicates that familiarity with the state of knowledge of food composition is necessary during the research planning stages. All too often, dietary intake records have been collected prior to determining the availability of the data for a food constituent. Research activities involving specific food components of current interest may occur far in advance of the acquisition of composition analyses, for example studies of dietary carotene intake. Constituents for which data have recently been requested are beta-carotene, total carotenes, retinol, Vitamin K, choline, chromium, iodine, selenium, total sugars, total refined (or added) sugars, individual sugars, all individual fatty acids, and dietary fibre. Values for some of these constituents are included in USDA *Handbook No. 8* revisions; however, for a large number of foods as consumed, data for these constituents are not available. These foods include many packaged, processed foods, fast foods, frozen ready-to-eat foods, and bakery products. Food items are added to the data base as requested by users; thus its composition does reflect its users' requirements. The data base at Case Western Reserve University is being updated continuously, with annual updates made available to owners.

In response to data-base owners' inquiries about updating the food table themselves, it is recommended that they use the recipe file software to accommodate those food items specifically for their use. This avoids the problem of overwriting their data in the food table as they receive updated versions.

It is important that users know how unknown nutrient values are accommodated in summaries. The summary reports a flagged nutrient sum if there is a missing nutrient value for one or more food items included for analysis. Additionally, users may be provided with an output format which indicates missing nutrient values for specific foods (fig. 1).

The degree of precision achieved with calculations of summaries is dependent on the precision of the nutrient values stored in the food table and the computer on which the programs are run. Guidelines for precision provided by USDA are used for inclusion of data from other sources. As smaller data bases are used on smaller computers, the degree of precision decreases. At the same time, as fewer food items are included in a data base, more substitutions must be made, thereby reducing the specificity of the reports for dietary intakes [10]. Such data bases may not be satisfactory for individual dietary records or as a general-purpose tool. These limitations must be understood and used for appropriate applications.

Inquiries have been received about the use of retention factors. The HVH-CWRU Nutrient Data Base does not incorporate retention factors and the issue is handled somewhat ambiguously. For most purposes, it is possible to select edible portions of foods as consumed. The recipe file system allows for coding of the edible portion of each ingredient by using yield factors for preparation and cooking losses. However,

Fig. 1. Report format for identification of missing nutrient data

ID:	NTN 343 01	19 yr female
Date:	Fri., 1 Feb 1985	Ideal wt: 52 kg
Name:	Sample meal	Ht: 165 cm Wt: 57 kg

		Kcal	T-Pro (g)	T-Fat (g)	Na (mg)	B₆ (µg)	Copp (mg)
12:00 Lunch at company cafeteria							
1 cup	Split pea soup, canned, diluted with equal volume water	184	10.0	4.3	975	66.15	0.3577
1 medium	Lettuce, iceberg, raw, leaf	3	0.2	0.0	2	11.00	0.0074
1 number	Peach halves, canned, solid and liquid, heavy syrup	60	0.4	0.1	5	15.39	0.0413
0.5 cup	Cottage cheese, creamed, large curd	116	14.1	5.1	455	75.38	
1 avg.	Corn meal muffin	126	2.8	4.0	192		
1 tsp.	Margarine, regular, soft, unspecified oils	33	0.0	3.7	50	0.368	
0.5 cup	Custard, baked	152	7.2	7.3	105		
1 cup	Whole milk, fluid, 3.3% fat, fort. with D	150	8.0	8.1	120	102.5	
1 small	Apples with skin, raw, EP, approx. 3 lb	54	0.2	0.3	0	44.16	0.0377

Nutrient summary			RDA	% cal.	Lytes
Kilocalories	878.0		44%		
Total protein	42.8 g		103%	20	
Total fat	32.9 g			34	
Sodium	1,903.0 µgᵃ		A		82.8 mEq
Pyridoxal B₆	314.9 µgᵃ		16%		
Copper	0.4441 mgᵃ		<A		

a. There is an unknown value for one or more food items.

this is not possible for recipes of most baked products, as many ingredients are coded in the raw, uncooked form, e.g. flour.

Another frequent request is for a data base of usual or common portion sizes. No attempts have been made to add these to the food table; however, this feature would be greatly appreciated for food-service and restaurant menu analysis. Planning menus or diets which meet specified nutrient requirements is an expressed need by a few users. The development of this capability is at present being explored as part of a clinical research study.

Ease of Access and Turn-around Time

The two requisites for ease of access and fast turn-around are a readily available computer and user-friendly, interactive processing and query procedures. The UNIX operating system affords an optimal time-sharing environment for direct access by telephone. Response time is fast and ideal for clinical and day-to-day use in practice or educational settings. Coding is still time-consuming and the most costly activity associated with analysis; it also requires knowledge of foods and nutrient composition.

Smaller data bases with fewer food items require greater coding knowledge, as more judgements regarding substitutions must be made than for larger data bases containing more foods and more adequate descriptive attributes.

Interactive on-line coding is desirable and would be more error-free; this will be implemented as computer time becomes less expensive. Response time, however, must be very short to alleviate boredom and tedium while entering data. For large data sets, batch processing with off-line data entry is more practical and less costly.

Report Forms

The most frequent reports requested are listings of foods with summaries of nutrient content and comparison with RDAs. Printed, easily readable formats are provided for verification and are most often used for clinical practice and patient education. Machine-readable formats must be used for computerized statistical analysis. Summaries of nutrient values or percentages of RDA or energy distribution may be aggregated by meal, time, place, food group, or any other set of descriptive features. Nutrients may be reported in ratios, the most common being nutrient/kilocalorie and polyunsaturated/saturated fatty acids. All reports may be produced in printed table or machine-readable formats. Under development at the present time are procedures for directing nutrient analysis output into an existing data-base management system. This will facilitate the generation of almost any kind of desired report.

Many nutrient data users have requested that reports include intake evaluations with recommendations for dietary improvement. The provision of these evaluations is not in accordance with the present policy of providing analyses for professionals' interpretation and use.

Cost

From the data-base user's point of view the costs are high, while from the manager's point of view charges seem never to be high enough to support continuing maintenance. This includes all of the related activities involved in understanding users' needs, explaining how their expectations can or cannot be met, and the inevitable hand-holding throughout the process as users and the data-base team are educated. The costs to users of the data-base system comprised two major components: (a) the costs of development and maintenance of the food table and software and (b) the computer costs. As hardware has become less expensive and more efficient, development and maintenance activities represent the major portion of expenses.

Justification for the costs of using nutrient data-base systems for clinical research and applications has been difficult, owing to concern about the reliability of food composition data and the problems of obtaining food consumption information.

CONCLUSIONS

Managing a large data-base system is challenging and, at the same time, frustrating. External factors contributing to this ambivalence include the dynamic nature of the food supply and the proliferation of research findings implicating food, nutrient, diet, and health relationships. Within the relatively small community directly associated with accumulating and disseminating food composition information in a useful form, the issues are recognized and are being discussed [15]. Many user's needs, as now expressed, seem to require knowledge and capabilities beyond the present ability to supply complete and timely information. Maintaining close communication and continuing dialogue between users and providers of data is critical to achieving a consensus for producing reliable, accurate, and useful reports with consistency and comparability.

REFERENCES

1. American Heart Association, Northeast Ohio Affiliate, *Cooking Without Your Salt Shaker* (American Heart Association, Dallas, Texas, 1978).
2. L. Bardach, J. Geduldig, and I. D. Mandel, "Dietary Patterns in Caries Resistant vs. Caries Susceptible Adults," *J. Dent. Res.*, 59 (special issue A): 508 (1982), abst.
3. H. T. Bebb, H. B. Houser, J. C. Witschi, and A. S. Littel, "Nutritive Content of the Usual Diets of Eighty-two Men, *J. Am. Diet. Assoc.*, 61: 407–415 (1972).
4. H. T. Bebb, H. B. Houser, J. C. Witschi, A. S. Littel, and R. K. Fuller, "Calorie and Nutrient Contribution of Alcoholic Beverages to the Usual Diets of 155 Adults, *A.J.C.N.*, 24: 1042–1052 (1971).
5. P. J. Garry, J. S. Goodwin, and W. C. Hunt, "Nutritional Status in a Healthy Elderly Population: Riboflavin," *A.J.C.N.*, 36: 902–909 (1982).
6. P. J. Garry, J. S. Goodwin, W. C. Hunt, and B. A. Gilbert, "Nutritional Status in a Healthy Elderly Population: Vitamin C," *A.J.C.N.*, 36: 332–339 (1982).
7. P. J. Garry, J. S. Goodwin, W. C. Hunt, E. M. Hooper, and A. G. Leonard, "Nutritional Status in a Healthy Elderly Population: Dietary and Supplemental Intakes," *A.J.C.N.*, 36: 319–331 (1982).
8. M. H. R. Goodloe, G. Waters, C. G. Quinlan, and J. G. Barrow, "Studies in Atherosclerosis. IV. Electronic Computers in the Processing of Dietary Data," *A.J.C.N.*, 13: 304–310 (1963).
9. O. B. Hayes, S. Abraham, and C. A. Caceres, "Computers in Epidemiologic Studies," *J. Am. Diet. Assoc.*, 44: 456–460 (1964).
10. L. W. Hoover, "The Problems of Using Small Nutrient Data Bases," *Proceedings of Eighth National Nutrient Data Bank Conference, July 25–27, 1983* (US Department of Commerce National Technical Information Service, Springfield, Va., 1983).
11. H. B. Houser, "Nutrient Data Banks – the State of the Art," Introduction in G. Petot and H. B. Houser, eds., *Proceedings of Fourth National Nutrient Data Bank Conference, April 23–24, Case Western Reserve University, Cleveland, Ohio, 1979* (Case Western Reserve University, Cleveland, Ohio, 1987).
12. H. B. Houser, A. I. Sorensen, A. S. Littel, and J. C. Vandervoort, "Dietary Intake of Non-hospitalized Persons with Multiple Sclerosis. I. Food Diary and Coding Methods," *J. Am. Diet. Assoc.*, 54: 391–397 (1969).
13. B. P. Klein, "Nutrient Analysis Information for Decision Making," *Food Tech.*, December 1983: 37–54.
14. L. J. Lillien, A. M. Huber, and M. M. Rajala, "Diet and Ethanol Intake during Pregnancy," *J. Am. Diet. Assoc.*, 81: 252–257 (1982).
15. G. J. Petot, "Commonalities and Differences: Are We All Speaking the Same Language?" *Proceedings of Eighth Nutrient Data Bank Conference, July 25–27, 1983* (US Department of Commerce National Technical Information Service, Springfield, Va., 1983).
16. G. J. Petot, "Nutrient Database Systems in Patient Care and Research: Clinical Applications," *Proceedings of Seventh Annual Symposium on Computer Applications in Medical Care, October 23–26, 1983, Baltimore* (IEEE Computer Society Press, Silver Springs, Md., 1983).
17. G. J. Petot, H. B. Houser, and R. V. Uhrich, "An On-line Nutrition Information System for the Clinical Dietitian," *Proceedings of Fourth Annual Symposium on Computer Applications in Medical Care, November 2–5, 1980, Washington, D.C.* (IEEE Computer Society, Long Beach, Ca., 1980).

18. G. J. Petot, A. Whitney and H. B. Houser, *Index of Nutritional Quality of Foods*, contract with Food and Nutrition Service, USDA. March 1979.
19. G. J. Petot, A. Whitney and H. B. Houser, *Nutrient Composition of Foods Commonly Available in the Marketplace*, contract with Federal Trade Commission. August 1978.
20. G. J. Petot, A. Whitney, and H. B. Houser, *Percentage of U.S. RDA in Frequently Used Serving Sizes and in 100 Kilocalorie Portions of Foods Commonly Available*, contract with Food and Nutrition Service, USDA. December 1979.
21. E. M. Thompson and H. Tucker, "Computers in Dietary Studies," *J. Am. Diet. Assoc.*, 40: 308–312 (1962).
22. S. Wassertheil-Smoller, H. G. Langford, M. D. Beaufox, et al., "Effective Dietary Intervention in Hypertension: Sodium Restriction and Weight Reduction," *J. Am. Diet. Assoc.*, 85: 423–430 (1985).
23. S. Wassertheil-Smoller, S. L. Romney, J. Wylie-Rosett, et al., "Dietary Vitamin C and Uterine Cervical Dysplasia," *Am. J. Epidem.*, 114: 714–724 (1981).
24. B. K. Watt and A. L. Merrill, "Composition of Foods: Raw, Processed, Prepared," *Agriculture Handbook No. 8* (Science and Education Administration, USDA, Washington, D.C., 1963 and revisions).
25. J. B. Wellisch, S. D. Hanes, L. A. Jordan, K. M. Maurer, and J. A. Vermeersch, *The National Evaluation of School Nutrition Programs: Final Report*, vols. 1 and 2 (System Development Corporation, Santa Monica, Ca., 1983).
26. J. C. Witschi, A. S. Littel, H. B. Houser, and A. I. Sorensen, "Dietary Intake of Non-hospitalized Persons with Multiple Sclerosis. II. Nutrient Intake for One Year Compared with Recommended Allowances," *J. Am. Diet. Assoc.*, 56: 203–211 (1970).

International Food Composition Data

14
Nutrient Intake Data Calculated Using Food Composition Tables: Factors Affecting Accuracy

HERNANDO FLORES and MARIA A.S.C. COELHO

Laboratory of Nutritional Biochemistry, Department of Nutrition, Federal University of Pernambuco, Cidade Universitaria, Brazil

INTRODUCTION

Specific nutrient deficiencies continue to be one of the world's major public health problems, especially in underdeveloped countries. In terms of number of individuals affected and geographical distribution, vitamin A and iron deficiencies are among the most prominent [25]. One very serious drawback for the design and implementation of nutrition intervention programmes is the inadequacy of dietary information, almost always plagued by the spectre of inaccuracy [21]. Thus, it is not uncommon to find reports describing high iron intakes in areas where iron-deficiency anaemia appears with undesirably high prevalence [2, 9, 11], or extremely low vitamin A intakes which are not accompanied by a compatibly high prevalence of eye lesions [2, 6, 9, 11]. Similar problems exist when trying to establish correlations between the intake of other nutrients and related clinical or biochemical indicators [3, 12, 18, 20, 21].

The accuracy of nutrient information depends on the methods of collecting and handling the data. Some of these have been examined [3, 8, 14, 20, 21] but, in general, attempts to reconcile dietary and biochemical or clinical information from nutrition surveys are still needed. The present study was designed to evaluate the relative contribution to the inaccuracy of dietary information of both regional differences in the nutrient composition of foods and the differences between those values obtained by calculation and those obtained by direct analysis of foods as eaten.

MATERIALS AND METHODS

Three recipes for the dishes most frequently consumed by the population in Northeast Brazil were selected for this study. These recipes are practically standard and appear with very little variation among users [19]; they can thus be considered representative of local alimentary practices. The proportion of raw ingredients and the per cent composition as actually eaten are shown in table 1. The culinary preparations, also in accordance with local practices, were as follows:

Table 1. Composition of regional dishes

Ingredients	Raw (%)	Cooked (% in serving)
Feijoada		
Beans (*mulatinho*)	39.5	21.1
Beef, dried	19.8	24.2
Bacon	4.9	10.1
Okra	2.0	2.7
Pumpkin	15.8	21.7
Green herbs[a]	14.8	18.9
Salt	1.0	–
Sarapatel		
Pork blood	39.5	39.5
Pork liver	12.8	12.8
Pork heart	12.9	12.9
Chard	11.9	11.8
Green herbs[a]	21.8	21.7
Salt	1.0	–
Cozido		
Beef	42.6	24.0
Pumpkin	9.1	8.9
Sweet potatoes	9.1	10.9
Potatoes	9.1	9.1
Plantain	7.3	11.6
Wild cabbage	3.7	7.4
Green herbs[a]	18.8	15.0
Cassava flour	–	5.1
Salt	1.0	–

a. A mixture, in equal parts, of green pepper, coriander, green onions, tomatoes, and onions.

- *Feijoada*: the beans (*mulatinho* type) were soaked in water for one hour, after which the beef and bacon were added. The mixture was boiled for two hours. Then the vegetables were added for a final boiling for 30 minutes.
- *Sarapatel de porco*: the pork blood and viscera were cooked in salted water, cut in small pieces, and boiled for one hour with the vegetables.
- *Cozido*: the meat and aromatic herbs were boiled in water for two hours. When these were nearly cooked, the vegetables were added. Before serving, the solids and liquid were separated, the latter to be mixed with cassava flour.

Individual servings of each preparation, in accordance with local uses [19], were duly homogenized with a blender, and appropriate aliquots of the homogenates were taken for the analysis of moisture, ash [10], fibre [24], ether extract [16], protein [15], calcium [4], phosphate [5], iron [10], vitamin A and carotenoids [1], and vitamin C [23]. The minerals were analysed in aliquots of the ashes. Carbohydrates were calculated by difference. The net weight of raw and cooked ingredients and the relative contribution of the latter in individual servings were recorded. All assays were run in duplicate, using appropriate standards. Differences greater than 5 per cent between duplicates were considered unacceptable. All ingredients of the recipes were also analysed individually. The nutrient composition of the recipes was calculated using the values of

138

the Food Composition Table of INCAP-ICCND [17] and those obtained from our analysis of the individual raw ingredients. Hence, two estimates of the nutrient composition of each recipe were made, along with direct analysis of appropriate aliquots of each dish. All reagents were analytical grade.

RESULTS

Table 2 permits comparison of the values of the INCAP table [17] with those obtained at our laboratory for the proximal composition, plus the iron, vitamin A, and vitamin C content of the 22 ingredients of the 3 dishes selected for this study. The values for fibre, calcium, and phosphate were not included in the table, as they add little to the objective of this work. It is readily apparent that agreement within a margin of ± 20 per cent was obtained only in about one-third of the total number of analyses performed. Results were below 80 per cent of the table value in 38 per cent of the observations, while 28 per cent presented with differences of more than 20 per cent above the figure in the table.

In the case of iron and vitamin A, the tendency was for the INCAP table values to grossly overestimate the content of the foodstuffs analysed. Especially remarkable was the case of iron, where 90 per cent of our values were well below 80 per cent of the figure presented by the INCAP table. Most of the major differences in protein content were in foodstuffs that are unimportant as sources of this nutrient, like coriander, pepper, and onions.

Table 3 was constructed to show the practical implications of the differences in the nutrient compositions described above. A previous survey [2], carried out in a country village, was used as a source of data to recalculate nutrient intake using the values obtained at our laboratory. Protein consumption shows very little difference, as could be expected from table 2. Vitamin A intake, on the other hand, seems to have been slightly underestimated, and that of iron grossly overestimated, when the INCAP table values were used.

The problem of "foods as eaten" was approached by comparing the results of direct analysis of the dish with the nutrient composition calculated using the values in the INCAP table or the the values obtained by local analyses (table 2). The changes in the relative proportion of the ingredients after cooking were taken into account by directly weighing the ingredients before and after culinary processing.

Figure 1 shows that up to 22-fold differences could be found when comparing the nutrient composition of the food as eaten with that calculated from the composition of the raw ingredients ("recipe calculation"). As could be expected from the degree of "complexity" of the dishes, *feijoada* was the one containing the largest differences.

DISCUSSION

It should first be emphasized that our aim was not to construct a Food Composition Table for local foodstuffs but, rather, to evaluate the degree of accuracy that can be expected from nutrient-intake data calculated from a Food Composition Table for Use in Latin America (i.e. recommended for this purpose).

It is apparent that differences of up to one order of magnitude can be found between the results of our local analysis and the values for some nutrients given by the INCAP

Table 2. Nutrient composition of local foodstuffs, Recife

Foodstuff	INCAP table no.	Moisture (%)	Protein (%)	Fat (%)	Carbo-hydrate (%)	Ash (%)	Iron (mg/100 g)	Vitamin A (µg/100 g)	Vitamin C (mg/100 g)
Pork blood	549	77	17.53	0.12	3.56	1.79	14.70	–	1.31
Pork liver	552	56	26.98	2.30	13.32	1.40	1.40	4,441	40.88
Pork heart	540	68	20.37	3.92	9.58	2.73	14.00	–	5.06
Coriander	143	90	1.10	2.75	3.14	2.99	11.60	1,291	129.00
Green onions	142	89	1.20	3.55	4.63	0.82	0.31	268	47.81
Green pepper	80	92	0.24	3.33	3.60	0.53	0.22	270	190.50
Tomato	271	95	0.49	0.55	2.92	0.65	0.09	368	16.69
Onion	137	85	0.21	2.41	11.57	0.55	0.20	–	11.62
Corn flour	27	11	8.62	1.76	77.54	0.56	0	26	–
Beef, lean	579	63	21.93	6.25	7.61	1.12	0.06	–	2.02
Pumpkin	127	68	0.71	0.14	26.45	2.29	0.27	1,526	16.88
Sweet potato	108	64	1.57	0.10	31.02	0.81	0.04	10	26.63
Sweet potato, yellow	107	–	–	–	–	–	–	–	–
Potatoes	242	86	0.71	0.11	11.63	1.27	0.34	0	32.00
Plantain	438	58	1.16	0.59	43.46	1.25	0.30	88	34.03
Cabbage, wild	149	90	4.63	0.17	3.19	1.32	1.44	–	190.69
Cassava flour	276	7	0.52	0.18	88.68	1.14	0.98	–	13.50
Beans, *mulatinho*	481	13	25.52	1.78	52.38	3.54	4.97	–	2.26
Beef, dried	582	28	56.80	16.32	17.07	15.95	6.90	–	1.13
Bacon	682	7	0.27	80.04	11.05	1.64	–	–	1.16
Anguria	–	91	1.35	0.05	4.74	0.41	0.34	–	36.38
Okra	252	83	1.61	0.06	12.85	1.21	0.56	14	27.06

Table 3. Estimated iron and vitamin A intakes using nutrient composition figures from INCAP table and analysis[a]

Nutrient	Table	Analysis
Protein g/d	20.9	21.7
Iron, mg/d	10.5	3.0
Vitamin A, μg/d		
(as retinol equivalents)	143.2	174.9

a. Food consumption data are from a survey in Agua Preta, Pernambuco [2].

table (the vitamin C content of green pepper determined here was 20 times higher than that given by the table). Of particular interest was the observation that nearly all the values for iron content were below 50 per cent of the value given by the INCAP table. It can readily be appreciated that the use of the table leads to an overestimation of iron consumption, and to reconciling the dietary data with the high prevalence of iron-deficiency anaemia found in the region [2, 11, 22]. It is also interesting to note that no value for the iron content of the 20 foodstuffs analysed here fell within 80–120 per cent of the value in the INCAP table. Equally interesting is the fact that the protein content of the so-called "sources of protein" showed little difference between the two values compared here. This might be the starting-point for suggesting that nutrient composition data could be divided into two categories: those of nutrients that show a high variation – probably attributable to regional differences (soil, climate, season, species) – and those of nutrients in some foodstuffs that show very little variation, probably insignificant for dietary evaluation purposes. Minerals and some vitamins are likely examples of the first category, while protein – being a compulsory component of foodstuffs derived from animal or plant tissues – could be a good example of the second category. Appropriate software for identifying the members of each category could be easily developed, and there are probably enough data available from various food composition tables to be used for this purpose.

Vitamin A nutriture constitutes a problem that should be looked at – in our region – from another angle. It is apparent that the difference in the figures for consumption resulting from the use of the INCAP table and our results is not enough to explain a lack of prevalence of severe signs of vitamin A deficiency [2, 6, 7, 12] which is not compatible with the very low vitamin A intakes reported in several surveys [2, 9, 12]. One possible explanation might be that some regional fruits, with a very high carotene and carotenoids content, are consumed by the population but not reported in the surveys. We have observed that the fact that some of these fruits are not actually "bought" may lead the population not to consider them as "foods." Thus, a significant contribution to vitamin A intake may have been overlooked in the past.

The problem of regional differences in nutrient composition – and the difficulties generated by the use of food consumption tables which are, most of the time, inadequate for specific situations – is well known. Our data have only shown what the practical implication of this may be, and one way to reconcile dietary data with other indicators of the nutritional status. Our data on "dish-nutrient composition" (fig. 1) shows another very serious drawback in the analysis of survey data with the aid of food composition tables: the so-called "foods-as-eaten" problem. In theory, there has never been any reason to consider as reliable "recipe composition data," i.e.

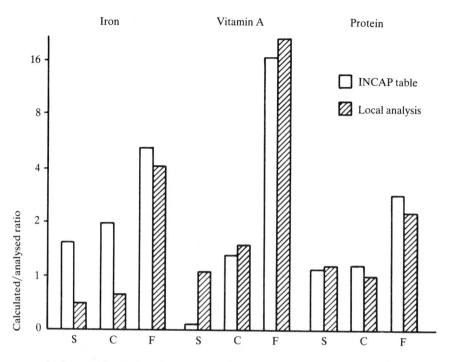

S: *Sarapatel* (pork viscera)
C: *Cozido* (meat and vegetables)
F: *Feijoada* (beans, pork and beef, vegetables)

Fig. 1. Dish composition: calculated *v.* analysed values.

the compound nutrient composition of a dish, obtained by addition of the contribution of each single (raw) ingredient. This approach ignores liability to heat of a great many nutrients, and losses that may result from chemical reactions as a consequence of the interaction between ingredients which are "incubated" for variable periods at 100°C or more. From a chemical view, *feijoada* was the dish to undergo the most drastic treatment (see recipe) and, in keeping with this, losses of 66 to 95 per cent were observed for protein and vitamin A respectively, regardless of the nutrient composition value used for the "recipe calculation." It was beyond the scope of this work to determine the actual causes for these losses, and we are first to admit that that of protein was the most intriguing. But the obvious conclusion is that food composition tables cannot continue to be used, without restriction, for calculating nutrient intakes.

The number of foodstuffs and nutrients analysed here was very modest. One should bear in mind, however, that sweet potatoes, cassava flour, sugar, beans, and a little meat account for more than 70 per cent of the daily food intake of the underprivileged in this region [2], as well as in most of the rest of the country. This is what we consider to be "alimentary monotony." Regarding nutrients, our contention is that emphasis should always be given to those capable of generating, by deficient or excessive intake, public health problems. These nutrients would include protein, vitamin A, and iron; the list would vary according to region, but would certainly be limited in each case. It is becoming increasingly important to count on reliable sources of accurate data for nutrient intake evaluation in connection with a number of nutrition-related diseases. Our data show that food composition tables do not meet accuracy requirements when the analysis involves foods that are not consumed raw, and where the presence and amount of nutrients in foodstuffs are very dependent on local conditions, mainly soil composition. This work also shows that dietary data can be reconciled with related clinical and biochemical indicators.

REFERENCES

1. C. R. C. Araujo and H. Flores, "Improved Spectrophotometric Vitamin A Assay," *Clin. Chem.*, 24: 386 (1978).
2. M. Batista, N. Chaves, R. M. Varela, M. H. M. Martins, A. C. Salzano, M. O. Bazante, S. M. G. Teixeira, G. J. C. Lima, F. M. Reis, G. C. Martins, and E. R. Linhares, "Inquérito em área urbana da Zona Mata do Nordeste Brasileiro. Agua Preta, Pernambuco," *O Hospital*, 79: 139–155 (1971).
3. J. P. Church, J. T. Indd, C. W. Young, J. L. Kelsey, and W. W. Kim, "Relationships among Dietary Constituents and Specific Serum Clinical Components of Subjects Eating Self-selected Diets," *A.J.C.N.*, 40 (Suppl.): 1138–1144 (1984).
4. P. V. A. S. Ferro and A. N. Ham, "Colorimetric Determination of Calcium by Chloronilic Acid. II. A Semimicro Method with Reduced Precipitation Time," *Am. J. Clin. Pathol.*, 28: 689–692 (1957).
5. C. H. Fiske and V. Subarow, "The Colorimetric Determination of Phosphorus," *J. Biol. Chem.*, 66: 375–400 (1925).
6. H. Flores, C. R. C. Araujo, F. A. C. S. Campos, and B. Underwood, "Importance of the Early Diagnosis of Vitamin A Deficiency at the Epidemiological Level," *Int. J. Vit. Nutr. Res.*, 24 (Suppl.): 23–34 (1983).
7. H. Flores and C. R. C. Araujo, "Liver Levels of Retinol in Unselected Necropsy Specimens: A Prevalence Survey of Vitamin A Deficiency in Recife, Brazil," *A.J.C.N.*, 40: 146–152 (1984).
8. C. G. Humble, J. M. Samet,. and B. E. Shipper, "The Impact of Revisions in Vitamin A Content Data on Estimates of Nutrient Intake," *Nutr. Res.*, 5: 175–179 (1985).
9. Fundação Instituto Brasileiro de Geografia e Estatística (FIBGE), *Consumo alimentar: Despesas das famílias: Tabelas seleccionadas* (FIBGE, Rio de Janeiro, 1978).
10. Instituto Adolfo Lutz, *Normas analíticas do Instituto Adolfo Lutz*, vol. 1: *Métodos químicos e físicos para análise de alimentos* (Instituto Adolfo Lutz, São Paulo, 1976), pp. 17–29.
11. Interdepartmental Committee on Nutrition for National Development (ICNND), *Northeast Brazil; Nutrition Survey, May, 1963* (ICNND, Washington, D.C., 1965).
12. Instituto de Nutrição da Universidade Federal de Pernambuco (INUFPE), *Hipotivaminose A no Brasil* (Ministério da Saude/INAN, Brasilia, 1977).
13. G. L. Kerr, E. S. Less, N. K. M. Lam, R. J. Lorinor, E. Randall, R. N. Forthkofer, M. A. Davis, and S. M. Magnetti, "Relationships between Dietary and Biochemical Measures of Nutritional Status in HANES I Data," *A.J.C.N.*, 35: 294–308 (1982).
14. W. W. Kim, W. Mertz, J. T. Judd, M. W. Marshall, J. L. Kelsay, and E. S. Prather, "Effect of Making Duplicate Food Collections on Nutrient Intakes Calculated from Diet Records," *A.J.C.N.*, 40 (Suppl.): 1333–1337 (1984).
15. C. A. Lang, "The Microdetermination of Kjeldahl Nitrogen in Biological Materials," *Anal. Chem.*, 30: 1692–1694 (1958).
16. R. Lees, *Food Analysis: Analytical and Quality Control Methods for Food Manufacturer and Buyer*, 3rd ed. (Leonard Hill Books, London, 1975), p. 91.
17. W. T. W. Leung and M. Flores, *Tabla de composición de alimentos para uso en América Latina* (Instituto de Nutrición de Centroamerica y Panama (INCAP) and Interdepartmental Committee on Nutrition for National Defense (ICNND), Guatemala City, 1961).

18. G. V. Mann, G. Pearson, T. Gordon, and T. R. Dawber, "Diet and Cardiovascular Disease in the Framingham Study," *A.J.C.N.*, 11: 220–225 (1962).
19. M. H. S. Martins and T. Karreco, "Valor nutritivo de alguns pratos regionais," *Boletim do Instituto de Nutrição da Universidade Federal de Pernambuco* (Recife), 2: 11–129 (1972).
20. C. W. Miles, J. S. Collins, J. T. Holbrook, K. Y. Patterson, and C. E. Bodwell, "Iron Intake and Status of Men and Women Consuming Self-selected Diets," *A.J.C.N.*, 40 (Suppl.): 1393–1396 (1984).
21. A. A. Rider, B. M. Calkins, R. S. Arthur, and P. P. Nair, "Diet, Nutrition Intake and Metabolism in Populations at High and Low Risk of Colon Cancer. Concordance of Nutrient Information Obtained by Different Methods," *A.J.C.N.*, 40 (Suppl.): 906–913 (1984).
22. A. C. Salzano, M. Batista, H. Flores, and C. L. A. Calado, "Prevalência de anemia no ciclo gestacional em dois estados do nordeste brasileiro, Pernambuco e Paraiba," *Rev. Bras. Pesq. Med. Biol.*, 13: 211–214 (1980).
23. R. Strohecker and H. M. Henning, "Vitamin C Determination," *Vitamin Assay* (Verlag Chemie, Berlin, 1965), p. 243.
24. A. L. Winton and K. B. Winton, *Análisis de alimentos* (Ed. Continental, Mexico City, 1957), pp. 75–79.
25. World Health Organization, *Vitamin A Deficiency and Xerophthalmia*, Technical Report Series, no. 590 (WHO, Geneva, 1976).

15
The Status of Food Composition Data in Asia

AREE VALYASEVI

Institute of Nutrition, Mahidol University, Bangkok, Thailand

INTRODUCTION

The First ASIAFOODS Conference was held from 17 to 21 September 1984 [2]. A total of 22 representatives from 12 Asian countries, as well as 17 resource persons and observers, met in Bangkok to review the current status of food composition data among Asian countries and to discuss the needs to improve food composition data generation, compilation, and use within Asia. An interim executive committee was appointed, with a representative each from Nepal, Sri Lanka, Indonesia, the Philippines, the Republic of Korea, and Japan; Dr. Aree Valyasevi of Thailand was appointed as the chairman of the committee. The executive committee subsequently met in Manila on 18–19 February 1985 [1] to: compile a regional survey of needs relating to food composition data; develop mechanisms for collaboration both within and outside the region; adopt statutes for the ASIAFOODS organization; and develop a five-year action plan, as well as proposals to obtain the assistance required to accomplish the plan.

Because Asia is the largest continent, with a population of over 2 billion, ASIA-FOODS has agreed to divide it into three subregions. Those countries invited to partici-pate in the initial ASIAFOODS meeting were distributed among the three subregions as follows:[1]

1. South Asia: Bangladesh, China, India, Nepal, and Sri Lanka.
2. South-East Asia: Brunei, Burma, Indonesia, Malaysia, the Philippines, Singapore, and Thailand.
3. East Asia: Japan, Papua New Guinea, the Republic of Korea, and Taiwan.

These divisions of ASIAFOODS attempt to reflect not only geographical proximity within the Asian continent, but also similarities between the climates, agricultures, and consequently food availability and dietary patterns. By this consolidition into

---

1. Representatives of Bangladesh, India, Brunei, and Singapore were unable to attend the First ASIAFOODS Conference.

145

similar subregions, data and analytical methodologies can be shared between the countries of the subregion and ASIAFOODS, resulting in the effective generation, compilation, and dissemination of high-quality food composition data.

FOOD AVAILABILITY

Considering the size of Asia and the consequent geographical and cultural diversity of the continent, these subdivisions are difficult to make, and all contain compromises with regard to the criteria of categorization employed. None the less, all current members of ASIAFOODS collectively developed this regionalization as a necessary administrative structure, and are satisfied with its arrangement.

With regard to foods produced and consumed within the member countries of ASIA-FOODS, there is, of course, tremendous diversity and variety. The range of available species of land and sea animals in addition to the cultivars of fruits and vegetables spanning the region is truly immense. Within each of the regions, however, there are sufficient similarities to allow for the systematic development, utilization, and sharing of necessary methologies.

The national surveys of foods produced, imported, and exported, as reported at the First ASIAFOODS Conference, indicate that the staples for the countries within the region are provided by a wide variety of indigenous cereals and tubers. These are supplemented by the importation of both indigenous and exotic foodstuffs.

For the region as a whole, wheat, maize, rice, dairy products, edible oils, frozen meats, and live animals seem to represent the predominant food imports. Exports consist of many of the imports listed above (excluding wheat and maize), with the addition of freshwater and salt-water fish, shellfish, tropical fruits, spices, coconut, and raw and refined sugar.

The subregion of East Asia leads ASIAFOODS in the production and consumption of processed or "manufactured" foods, and in their importation of the unprocessed agricultural products used in those foods. There is, as yet, no significant exportation from this subregion of either processed or unprocessed foods, although internally there is some international trade in these commodities.

The South-East subregion is characterized largely by the importation of basic food commodities for consumption as dietary staples, along with that of foods that are not widely produced in the region (predominantly dairy products). This is offset by the export of a wide variety of foods, including tropical fruits, seafoods, coconut, and cassava.

The countries of the East and South-East Asia subregions all depend upon rice as the universal staple, with the exception of Papua New Guinea, where the consumption of rice continues to increase. Each of these nations also has a proportionally large seacoast, and so seafoods represent a major part of the available food.

The South Asia subregion would have to be considered a slight net importer of foodstuffs, again with dietary staples being imported and tropical agricultural products – notably fruit, spices, and tea – being exported. Some localities within this subregion suffer from a food deficit, and consequently the inhabitants rely upon wild plants for food on a seasonal or, in some cases, a continual basis. There exist throughout this subregion traditional food processing methods, in addition to rapidly developing modern food-processing technologies; all of these food categories are very much in need of nutrient analysis.

146

The South Asia subregion contains India and China (both vast countries), which, along with Nepal, Sri Lanka, and Bangladesh, share rice and wheat as the predominant staples. They both contain landlocked localities and areas with ready access to the sea. Although there are local variations, underlying similarities exist throughout China and India, and carry over to the countries adjacent to them.

The diversity of available foods within the ASIAFOODS region is perhaps exceeded only by the region's cultural diversity, which is expressed through the culinary and dietary traditions of the various peoples. Again, it is intended that the subregions capitalize upon the consistencies inherent within each subregion.

GENERATION OF FOOD COMPOSITION DATA

The pace of development of food composition analysis differs for each country within a subregion, as it does between the subregions themselves. Generally speaking, however, South Asia appears to be in a more initial stage of development and East Asia at a more advanced stage, with the South-East Asia subregion lying somewhere between. Several countries in East Asia, for example, have within the last few years published revisions of their national food composition tables, incorporating many new foods as well as analysing more nutrients such as amino acids and fatty acids. None the less, even in this subregion there remains a large deficiency in the number of nutrients analysed. Some vitamins, fatty acids and cholesterol, dietary fibre, and trace minerals are examples of nutrients that remain to be analysed in many countries. Indeed, Nepal and Sri Lanka in South Asia are obliged to use food composition tables from India, since their own national tables are extremely limited.

For each country in the region to develop its own food composition tables, revisions must be made in the method of generating data. Currently, methods of food analysis are not uniform and there is a general lack of standardization and quality control. Also, the systems or formats for data presentation are dissimilar, and this can affect the system of data generation. The data contained in any food composition data table have to be unquestionably reliable in terms of accuracy, not only for national domestic use, but also for use outside the country. This goal – a reliable and readily understandable food composition table – is within the mandate of ASIAFOODS and INFOODS.

The food-analysis methodologies that will be used by ASIAFOODS member laboratories can be developed and refined on a regional or subregional basis. This will provide for both regional standardization and significant savings, since each country will not need to develop, test, and refine each particular analytical methodology on its own.

Nutrients will be determined using the same methodology and analysing the same specific constituent, and they will be measured in the same units. Consequently, all food data generated within the ASIAFOODS region will be compatible and interchangeable; and food composition data generators, compilers, and users will have made the first major step towards international collaboration.

For the major of the countries in ASIAFOODS, food composition data are made available through published food composition tables which have been produced within the country. These national tables are then augmented by regional ones, which have been produced either by international agencies or the governments of developed countries. A few of the countries in ASIAFOODS must rely solely upon these foreign data tables, somehow cobbled together so as to satisfy the requirements of users within that country.

Even for those countries that have the capacity to produce their own national food tables, there are numerous problems associated with these tables as they currently exist – principally those of revising, updating, and formating data, and the lack of a standard system for presenting data. The cost and effort involved in editing and publishing a national food composition table rules out the possibility of publishing frequent revisions. This means that the tables are unable to reflect technological advances in food composition analysis or the additional data available through the analysis of new foods.

The recent striking advances made in nutrient analysis, i.e. vitamin A determination with high-performance liquid chromatography, casts doubt on much of the published food composition data as regards vitamin A. Furthermore, since the cost of updating current food composition tables can only be justified every 10 to 20 years or so, the time lag between the development of new analytical techniques and the publication of values for nutrients often undermines users' confidence in the tables. In countries where multiple food tables are in use, nutrient values for the same food often differ, again undermining the user's confidence in all the tables involved.

Currently, food composition tables must present all available food composition data that might be required by the broadest existing user group. The ease and cost of publication and distribution of tables could be improved if it were possible better to determine the needs of more specific user groups and to tailor more specialized tables to meet the demands of those groups only.

Countries that publish food composition tables (and here it should be noted that not all countries publish those data in tabular format) often design them according to fairly arbitrary national criteria. This means that the tables from different countries can be slightly or enormously different, in terms of the structural and visual organization and presentation of their food composition data. Beyond this, the data can differ in other technical ways which can make two tables completely incompatible or compatible only through cumbersome calculations. The frustration of this situation for those countries that must rely on a package of foreign food composition tables for national use does not need to be elaborated upon.

With data generation standardized, data compilation can also move ahead to the point where any food table in the ASIAFOODS region can be used by any nutrition professional anywhere else in the region with complete facility. Through INFOODS, this same consistency is to be extended worldwide. With the long-standing as well as the more recent migration of Asian populations around the world, and with the recent upsurge of interest in Asian foods and diets, there is an increasing global demand for reliable, easy-to-use, and understandable food composition tables relating to the foods of Asia. In addition, as Asian countries export more and more of their agricultural produce around the world, the need to know the exact nutrient composition of these foods will also increase. Information regarding the availability, generation, users, and uses of food composition data, as well as unmet needs and plans for the future on a country-by-country basis, is presented in table 1.

USERS AND USES OF FOOD COMPOSITION DATA

While there is a broad range of users of food composition data in each ASIAFOODS country, these fall into definite categories, as follows:

1. Nutrition researchers and health-care personnel, for nutrition and food consumption surveys.
2. Nutritionists, dieticians, and medical personnel, for diet prescription and nutrition management of in- and out-patients.
3. Nutrition educators and dieticians, for diet counselling and meal planning.
4. Nutritionists and food technologists in the food-processing industry.
5. Economists, planners, and consumer-protection personnel.

UNMET NEEDS

The perceived needs that the current food composition tables do not meet concern many of the processed foods that have only recently become widely available in ASIA-FOODS countries. Virtually all of the tables available list nutrients for uncooked food items, which means that since many nutrient values change with cooking these listings do not accurately reflect what is actually consumed. The vast majority of the population of many of these countries reside in rural areas rather than the urban centres. Since these people are usually poor, they are often compelled, at least on a seasonal basis, to consume what could be characterized as "wild" foods. There is essentially no data on these non-domesticated foods, and any efforts at nutrition intervention or education are handicapped without such knowledge.

One suggestion to come out of the most recent ASIAFOODS executive committee meeting was that there should be a differentiation, when analysing and presenting food composition data, between chemical content and physiologically or biologically active content, e.g. retinol and carotenoids. It has also been suggested that data on selected nutrients be related to the prevalent health problems in each country; for instance, the role that certain nutrients play in the prevalence of cancer, as well as degenerative diseases, is not clear. More attention should be paid to the functional aspects of the nutrients rather than their chemical composition.

Participants at both ASIAFOODS meetings expressed concern that there was a minimal awareness of needs relating to food composition data among planners and administrators in influential positions. These decision-makers need to be motivated and encouraged to support more vigorously those activities proposed by ASIAFOODS.

It was universally recognized that there is currently a shortage of qualified personnel to accomplish the goals of ASIAFOODS.

The recently launched ASEAN Food Habits Project will have a food composition component which will interact very closely with the food-composition data-generation activities of ASIAFOODS.

THE FUTURE OF ASIAFOODS

The initial steps to be taken by ASIAFOODS will be to establish national committees, a secretariat, and effective channels of communication between ASIAFOODS countries. The national committee will be the country's contact point with ASIAFOODS and will determine specific needs, mobilize available resources, and endeavour to establish an environment favourable to the development of good food composition data within the country. The ASIAFOODS secretariat will serve as the contact point for ASIAFOODS with the national committees. The overall activities of ASIAFOODS

Table 1. Available food composition data (FCD) in Asia countries

Food composition tables (FCT), latest date published	Data generators	Data users	Data uses	Unmet needs	Plans
Burma					
National FCT, 1969 FAO FCT for South-East Asia Nutritional Value Indian Foods	Dept. of Health Services Dept. of Medical Research	Nutrition/health professionals Primary health care planners	Nutrition assessment Formulation of weaning food Nutrition education	Not available	Not available
China					
National FCT, 1981	Nutrition and Food Hygiene Dept.	Hospital dieticians Nutrition workers Child-care centres Food manufacturers	Diet management of patients Diet planning	Unavailable data for cooked foods, snacks, and other common foods	Add new components to FCT
Indonesia					
National FCT, 1967	National Institute for Health Research and Development Nutrition Research and Development Center Other related institutes	Nutrition professionals Students Agricultural workers Central Bureau of Statistics	Food balance sheets Nutrition surveys	Incomplete FCT	Collect data on natural foods Re-analyse carotene, vitamin A, Fe, etc.
Japan					
National FCT, 1982 Selected component FCTs	Japan Food Research Laboratory National Institute of Health	Nutritionists Dieticians Students	Nutrition surveys	Selected nutrients related to health problems	Develop functional nutrition activities Develop system for updating FCT, etc.

150

Korea				
National FCT, 1981 Office of Rural Development National Institute of Health	Nutrition professionals Students Planners	National economic plans Food balance sheets Menu planning	Data for cooked foods, fast foods, infant formulae	Upgrade data compilation Edit RDA guide Develop education materials for all levels of knowledge
Malaysia				
National FCT, 1980 FAO FCT for East Asia Institute of Medical Research in Collaboration with Food Science and Technology Section of the Malaysian Agriculture Research and Development Institute (MARDI)	Nutritionists Dieticians	Foods consumption study	Insufficient information on local foods Incomplete nutrient contents Limited personnel, equipment, analysis methodology	Management of food composition data with computers
Nepal				
No National FCT Indian FCT FAO FCT for East Asia Central Food Research Laboratory – Ministry of Agriculture	Planners and decision-makers Nutrition/health professionals Home scientists Research workers	Formulate diets, weaning food	Lack of nutritive value and toxicity for wild foods	Develop sound food and nutrition policy Analyse available wild edible plants

Table 1 (*continued*)

Food composition tables (FCT), latest date published	Data generators	Data users	Data uses	Unmet needs	Plans
Papua New Guinea					
National FCT, 1978 Tables of Representative Values of Foods Commonly Used in Tropical Countries, 1962 South Pacific Foods, 1958 FCT in Pacific Islands	Papua New Guinea University of Technology University of Papua New Guinea Dept. of Primary Industry	Not available	Not available	Insufficient data for root crops, pandanus, tropical fish, and manufactured food	Upgrade food analysis equipment within the region
Philippines					
National FCT, 1980	Food and Nutrition Research Institute in conjunction with government and private research agencies	Nutrition/health professionals Food technologists Chemists Planners and economists	Nutrition information Nutrition assessment/ epidemiology Diet counselling, meal planning, plant and animal breeding Economic planning	Analysis methodology limited Insufficient nutrients analysed Insufficient analysis trials of each food item	Resolve sampling problems Determine which foods/nutrients to analyse Update FCT
Sri Lanka					
National FCT – 10% Sri Lankan foods and 90% Indian	Central Agricultural Research Institute Tea and Coconut Research Institute	Nutrition/health professionals Nutrition educators Nutrition and food planners	Not available	Complete Sri Lankan FCT Inadequate analytical equipment and facilities	Improve lab facilities Request assistance from outside agencies

	Food composition tables	Institutions	Users	Uses	Limitations	Future plans
		Ceylon Institute of Scientific and Industrial Research Medical Research Institute National Aquatic Resources Agency				
Taiwan	Composition of Foods Used in Taiwan, 1961 Table of Taiwan Food Composition, 1974 Table of Amino Acid Composition of Taiwan Foods, 1973 USDA FCT for East Asia FCT for Japan	Taiwan University Food industry Research and Development Institute	Dieticians University students	Not available	Inadequate and out-of-date FCT	Evaluate, re-analyse certain food items for FTC
Thailand	National FCT, 1978 Amino Acid Composition, 1984	Nutrition Division, Dept. of Health Institute of Nutrition, Mahidol University Division of Food Analysis, Dept. of Medical Science Division of Biological Science Thailand Institute of Scientific and Technical Research	Nutritionists Health research scientists Dieticians Food industrialists	Menu planning Product development Nutrition education Consumer protection labelling	FCT lacks information on some minerals, elements, fatty acids, sugars in fruits	National collaboration project to generate new data New analytical procedures Organize data in tables and computerize

will be organized and co-ordinated by the secretariat, which will be under the supervision of the chairperson of the executive committee.

The fundamental objective of ASIAFOODS is the sharing of scientific knowledge and international collaboration relating to food composition data. Useful and effective media for communicating this information throughout the region will need to be identified at the earliest opportunity.

The design and implementation of the activities necessary to realize these ASIAFOODS objectives can best be facilitated by meetings of the complete ASIAFOODS membership. Accordingly, a second ASIAFOODS conference will be convened, which will consist of a business meeting and a technical meeting.

The business meeting will address itself to ratifying the ASIAFOODS statutes as well as developing mechanisms of regional collaboration and liaison with INFOODS. A review of progress achieved with ASIAFOODS activities since the First ASIAFOODS Conference will also be made. Finally, the tasks and responsibilities of the ASIAFOODS executive committee and secretariat will be delineated and personnel will be selected to fill these positions.

The technical meeting will in turn be concerned with working through the INFOODS international guide to the development of food composition tables on a country-by-country basis, and to detail the needs of each country. Small technical meetings will be planned relating to food chemistry, information systems, and usage of food composition data. These meetings will be devoted to the detailed design of subsequent training workshops, and discussions and planning of training needs and system operation will take place. Training is visualized as occurring on three levels:

1. Inter-country training, whereby two or more countries themselves train laboratory workers in the updating, gathering, and handling of food composition data.
2. Regional training, intended for scientists at the supervisory level and to be offered at a specialized regional centre.
3. Inter-regional training, intended for high-level personnel who will visit advanced centres outside the region.

Subsequent to their initiation of these activities, a regional data centre will also be designed; particular attention will be paid to computer hardware selection and installation and the development of appropriate software, and to obtaining the services of a programmer. Arrangements will also be made for software support of INFOODS data interchange standards. There will also be initial training in the use of these facilities for specialists from the various countries of the region.

These activities form the core of the ASIAFOODS plan for the next five years. Drawn up by the executive committee, this plan intends each country in ASIAFOODS to be able to participate and benefit at a level appropriate to its technological capability and political commitment. Rapid progress has been made to date in establishing ASIAFOODS, and increased membership and enhanced scientific and technical collaboration throughout the region are anticipated.

The keen interest in ASIAFOODS and the objectives it is mandated to accomplish clearly reflect the present demand for more reliable and usable food composition data throughout the region. It is intended through the structure of ASIAFOODS to develop collectively a regional food composition data base which will satisfy this demand cost effectively. An international regional organization such as ASIAFOODS, in conjunction with INFOODS, provides a vehicle by which these various objectives can be realized.

154

ACKNOWLEDGEMENTS

The First ASIAFOODS Conference was partially supported by grants from IDRC (International Development Research Center, Canada) and the ASEAN Sub-committee on Protein.

The author would like to thank Mr Andrew Stuckey for his valuable assistance in the preparation of this manuscript.

REFERENCES

1. "Minutes of the ASIAFOODS Executive Committee Meeting, Manila, the Philippines, 18–19 February, 1985" (Institute of Nutrition, Mahidol University, Bangkok, 1985).
2. *Proceedings of the First ASIAFOODS Conference, Bangkok, Thailand, 17–21 September, 1984* (Institute of Nutrition, Mahidol University, Bangkok, 1985).

155

16
Food Composition Data in Sweden and the Nordic Countries

ÅKE BRUCE and LENA BERGSTROM

National Food Administration, Uppsala, Sweden

SWEDISH FOOD COMPOSITION TABLES

The National Food Administration is the central Swedish administrative authority on food matters. It belongs to the Ministry of Agriculture and has two main tasks: to protect the consumer against harmful foods and to help the consumer to evaluate and choose foodstuffs.

The National Institute of Public Health preceded the present administration and one of its tasks was the analysis of food items. These analyses together with data from the literature were compiled into a food composition table, *Fodoamnestabeller*, which was published by E. Abramson in 1947. Since then, these tables have been revised several times, the seventh edition being published in 1982 [1].

Analyses of foods for a comprehensive Swedish food composition table began in the early 1960s. The National Food Administration published such a table in 1978 and an abbreviated version in 1981. At present, the comprehensive table is being revised and extended to include more nutrients, and a new edition will be available in 1986 [10]. The abridged table was revised in 1984 [9].

SWEDISH NATIONAL NUTRIENT DATA BASES

In the early 1960s the National Institute of Public Health developed a nutrient data bank based mainly on the Abramson food composition table and data supplied by the food industry in Sweden. Since 1981, a new data bank has been prepared by the National Food Administration. This bank was first set up in order to expedite nutrient calculations in surveys, and began at the same time that the data collection for an extensive nutrition survey of children was under way in 1980–1981; it was first used for the nutrient analysis of the different food intake forms collected during this survey. The bank at present contains the nutritive values of 1,250 foods and about 1,200 recipes, including a large number of baby foods and certain catering products.

Table 1. Nutrients in the Swedish National Nutrient Data Base

Energy (kJ and kcal)[a]	Vitamin B_6
Water	Vitamin B_{12}
Ash	
Protein[a]	Calcium[a]
Fat[a]	Phosphorus
Carbohydrate[a]	Iron[a]
Alcohol	Magnesium
	Sodium
Retinol	Potassium
Active carotene	Zinc
Retinol equivalents[a]	
Vitamin D	Saturated fatty acids
Vitamin E total	Monoenes
Alpha-tocopherol	Polyenes
	Cholesterol
Ascorbic acid[a]	
Thiamine[a]	Monosaccharides
Riboflavin[a]	Disaccharides
Niacin	Sucrose
Niacin equivalents[a]	

a. Nutrients for catering products.

For each item, the values for the energy, water, ash, and 31 other nutrients are given. For catering products only 12 nutrients are included (table 1).

Of special interest are a number of codes containing averages for different food groups. There are, for example, nutritive values for fresh vegetables based on the average consumption of different vegetables. These aggregate codes are used in certain types of planning, and in the calculation of average consumption figures from official statistics, which is carried out in co-operation with the national Agricultural Market Board. These codes are also suitable for nutrient calculations of dietary history forms.

In recipe calculation, food yields and changes of the nutrient content of dishes have been considered. Food-yield factors are based on both American and Swedish values, while factors of vitamin losses are based on different published data. Analyses of nutrient changes during preparation of certain common Swedish dishes are now in progress at the Food Administration.

At present, information on several nutrients is being incorporated: data on additional vitamins, minerals, and trace elements, and also on different carbohydrate fractions (including added refined sugars), dietary fibre, and fatty acids. The enlarged bank will be ready for use in 1986.

The Food Administration's nutrient data-base system is adapted for the user. After a few hours of instruction, a person can enter a seven-day record and calculate the nutrient content. More advanced statistical treatment requires a few days' training. At present, either partial or complete copies of this nutrient data bank are being used by, for example, the Departments of Nutrition and of Home Economics Education in some Swedish universities, by the National Agricultural Market board, by the Material Administration of the Armed Forces, and by some county councils.

OTHER SWEDISH DATA BASES

During the last few years several nutrient data banks have been created and still more are planned. Today the Administration is aware of about 30 different systems.

The main applications of these systems are: nutrient calculation and analysis; recipe and product development; diet planning and analysis; menu planning; food production and control; and education and information.

The owners of nutrient data banks in Sweden represent the following categories: national authorities; regional and local authorities; universities and schools; food industries; wholesalers; and publishing and computer firms.

At the national level four national authorities are using nutrient data banks. The Administration can use the bank for nearly all purposes mentioned above but the most important task is nutrient analyses of large-scale nutrition surveys. The system used by the National Agricultural Market Board is applicable to nutrient calculation combined with food-consumption statistics and food-supply calculations.

The county councils of Sweden are responsible for health and welfare. Some of them are now using computers in the planning of cycle menus and of special diets and dietary analyses combined with patient counselling. Local authorities are also using their nutrient data banks for menu planning in different institutions, and for school lunches.

Five universities (in Gothenburg, Lund, Stockholm, Umea, and Uppsala) have access to nutrient data banks. All of them also have close connections with large research hospitals. For research, training, education, dietary counselling, planning of special diets, etc., several institutes and departments need nutrient data banks. Swedish university units that now use nutrient data bases include nutrition and dietetics, medicine, odontology, food science and economics, home economics, psychology, sociology, European ethnology, cultural anthropology, economic history, and economics.

Large food industries use the banks for nutrient analysis for labelling and product information. Recipe or product development with evaluation of proposed product changes is another application.

One Swedish wholesaler has a private nutrient data bank. The test kitchen of this firm uses the bank and provides many restaurants with nutrient-analysis and menu-planning services. For the present, the other wholesalers are buying these services.

The oldest of the private nutrient data-base systems is Diet and Nutrition Data. This company serves different user categories with nutrient calculations and analyses. Other systems are more specialized, serving restaurants, catering companies, hospitals, etc.

Nearly all the nutrient data banks in Sweden are under development; the systems are growing and becoming more sophisticated. At present, a certain co-ordination is desirable.

FOOD COMPOSITION TABLES IN OTHER NORDIC COUNTRIES

Denmark

Since 1973 two food composition tables compiled by Peder Helms have been used [2, 3]. Denmark's first official food composition tables were published in 1983, with a revised edition in spring 1986 [6].

Finland

In Finland there are no official food composition tables. However, those published by Osmo Turpeinen and Paavo Roine in 1952 have been updated regularly [11]. A comprehensive mineral and trace-element table of Finnish foods, edited by Pekka Koivistoinen, was published in 1980 [4] and two editions in Finnish by Pertti Varo came in 1980 and 1981 [12].

Data on dietary fibre and available carbohydrates in Finnish cereal products, vegetables, and fruits have been published by Pertti Varo et al. [13, 14].

At the Department of Food Chemistry and Technology of the University of Helsinki, vitamin E, carotenoids and fatty acids in Finnish foods are now being analysed.

Norway

The National Nutrition Council is responsible for the scientific content of the Norwegian official food composition tables; they are produced and published by the National Society for Nutrition and Health. The first edition of the tables was published in 1958 and the fifth in 1984 [5].

Iceland

Laboratories in Iceland analyse Icelandic foods and these analyses are published in, for example, the report series of the Agricultural Research Institute (RALA) [8]. Otherwise, British and American food composition tables are used.

NUTRIENT DATA BANKS IN THE OTHER NORDIC COUNTRIES

Denmark

In Denmark there are about 15 nutrient data-base systems. The official system was developed by the National Food Agency and is available on microcomputers in the agency and from the computer centres of the universities of Copenhagen and Aarhus. This system is mainly used for nutrition surveys and is now being completely revised. In the future it will include data on about 100 nutrients and many non-nutrients for approximately 10,000 food items.

Another agency system, DANKOST, suited for nutrition calculation and planning of diets, was developed for microcomputers in 1985.

The Agriculture Computer Centre and the Danish Academy of Engineering are other organizations which own nutrient data-base systems. Several hospitals have systems for menu planning and diet analysis, one of which was developed by the Danish Hospital Institute. At the moment hospitals in the Copenhagen area are developing computer-based menu planning systems. Educational systems for different school levels have been made available by the Ministry of Education. Small systems for weight reduction are also in use.

Table 2. Nutrient data banks in the Nordic countries: Finland, 1985

	The State Catering Centre	The Dept. of Nutrition, University of Helsinki	The Rehabilitation Research Centre	The National Public Health Institute
System owner				
Address	Boks 332 SF-00531 Helsinki 53 Finland	SF-00710 Helsinki Finland	Pettolavagen 3, SF-20720 ABO, Finland The Social Insurance Institute P.O. Box 640, SF-00101 Helsinki, Finland	Mannerheimvagen 166 SF-00280 Helsinki Finland
System name in operation since	1984	1972	New version, 1981	1985
Nutrient data sources	Turpheinen; Food Composition Tables, 1980; product information from food industry	Turpheinen; Food Composition Tables, 1980; Varo, Mineral Tables, 1982; other food composition tables – Swedish, 1978, British, 1967, Norwegian, 1977, German, 1981; literature; own analyses; product information	Food composition tables – Swedish, 1978, German, 1981, British, 1978, USA *Agriculture Handbook No. 8–I*; product information	Nordic food composition tables; other tables; own analyses; product information
Number of (a) foods, (b) recipes, (c) aggregated foods, a–b, a–c	(a) 360; (b) 78	(a) 450; (b) 400	(a) 500; (b) 400	(a) 450; (b) 700
Nutrients/food	8	60	70	65
Other components/food	Food cost data	No		Recipe procedures
Nutrition recommendation	No	No	RDA and Swedish Nutrition Recommendations	No

Recipe calculation	No	Yes	Yes	
Food yields	No	Yes	Yes	
Nutrient losses and gains; factors			Nutrient changes of every food item in the recipes	
System is used by	The staff of the Centre	Scientists and students of the department	Scientists, dietitians, etc., of the Centre or the Institute	The staff of the Institute
System is suited for	Nutrition surveys; diet planning; menu planning	Nutrition surveys; research; education	Nutrition surveys; research; diet planning	Nutrition surveys; research
System is used for	Calculation of nutrients; calculation of costs.	Calculation of nutrients	Calculation of nutrients	Calculation of nutrients
Computer		Borroughs 7800	IBM 30840	VAX
Program source				
Program language	APL	PASCAL	APL	PASCAL
Operating data-base management system			MVS/150	VAX/VMS 3, 6 VMS/RMS
Access		Modem and terminal	Modem and terminal	
For sale: Nutrient data-base system	No	No	No	No
Nutrient data base	No	No	No	No
Services				
Comments				

Table 3. Nutrient data banks in the Nordic countries: Norway, 1985

	The Section for Dietary Research University of Oslo	NUME Co., Ltd.	Ianum-Norli Publisher Co., Ltd	Ianum-Norli Publisher Co., Ltd.
System owner				
Address	Pb 11 17 Blindern N-0317 Oslo 3, Norway	Ivetenveien 152, Iveita N-0671 Oslo 6, Norway	Kr. Augustgt. 7A N-0164 Oslo 1, Norway	Kr. Augustgt. 7A N-0164 Oslo 1, Norway
System name in operation since	1969; new version, 1980	1985	1982	1984
Nutrient data sources	Food composition tables – Norwegian, 1984, Swedish, 1978, British, 1978, German, 1981, USA USDA *Agriculture Handbook No. 8,* 1963; product information from food industry	Food composition tables – Norwegian, 1984, Swedish, 1978, Danish, 1983, German, 1981, British, 1978, USA USDA *Agriculture Handbook No. 8-1*	Food composition table – Norwegian, 1977	Food composition table – Norwegian, 1977
Number of (a) foods, (b) recipes, (c) aggregated foods, a-b, a-c	(a) 800; (b) 1,000	(a-c) 1,600	(a) 700	(a) 256
Nutrients/food	28	150	12	11
Other components/food				Food cost data, standard and own measures
Nutrition recommendation	No	Norwegian Nutrition Recommendations	Danish Nutrition Recommendations	Norwegian Nutrition Recommendations
Recipe calculation				
Food yields	Yes			
Nutrient losses and gains; factors	No			

162

	The staff of the Section	Nutritionists and dietitians with high competence (*dietetiker*)	Stabekk, College of Education for Domestic Science, and Elvebakken School	Elementary school
System is used by	The staff of the Section	Nutritionists and dietitians with high competence (*dietetiker*)	Stabekk, College of Education for Domestic Science, and Elvebakken School	Elementary school
System is suited for	Nutrition surveys; research	Nutrition surveys; research; special diet planning; menu planning	Diet planning; menu planning; education	Education
System is used for	Calculation of nutrients	Calculation of nutrients	Calculation of nutrients; optimize recipes	Calculation of nutrients; calculation of costs
Computer	DEC 1099 (2CPU DEC 1090)	DEC	ABC 80	IIKI
Program source				
Program language	SIMULA	COBOL	BASIC	BASIC
Operating system	TOPS 7.02	MS DOS		
data-base management system	Own	Cogen		
Access	Modem, terminal			
For sale: Nutrient data-base system	No	Yes	Yes; on floppy disk	
Nutrient data base	No	Yes	Yes	
Services				
Comments				

Table 4. Nutrient data banks in the Nordic countries: Sweden, 1985

	The National Food Administration (NFA)	Gothenburg University Computer Centre	Stockholm County Council	Diet and Nutrition Data Co., Ltd.
System owner Contact person Address	Lena Bergstrom Box 622 S-751 26 Uppsala, Sweden	Monica Abrahamson Kapellgangen 5 Box 190 70 S-400 12 Gothenburg Sweden	Mariana Atterhult/Lena Kanstrom Box 9106 S-102 72 Stockholm Sweden	Lillemor Wallentinson Bradakersvagen 7 S-217 63 Malmö, Sweden
Telephone Telex	+46 18 17 55 00, 17 57 30 76121 SLVUPS S	+46 31 60 30 29	+46 8 737 25 00	+46 40 91 55 30
System name in operation since	KOSI 1981	FOODCOMP 1982	KIS-PLAN 1983	DIETIST 1969
Nutrient data sources	Food composition tables – Swedish, 1978, USA USDA *Agriculture Handbook No. 8*, 1963, British, 1978, German, 1981; own analyses; product information from food industry	Food composition tables – see NFA; own analyses; product information	Food composition tables – Swedish, 1978; product information	Food composition tables – Swedish, 1978; product information
Number of (a) foods, (b) recipes, (c) aggregated foods, a-b, a-c	(a) 850; (b) 400; (c) 130	(a) 840; (b-c) 510	(a) 2,000	(a) 1,500
Nutrients/food Other components/food	35	35	26 Food cost data	26 Food cost data
Nutrition recommendation	Swedish Nutrition Recommendations	No	Swedish Nutrition Recommendations; recommendations for special diets	Swedish Nutrition Recommendations; recommendations for special diets
Recipe calculation Food yields Nutrient losses and gains; factors	Yes Yes	Yes Yes	No No	No No

	The Administration; some departments of the universities of Uppsala and Stockholm	The Departments of Clinical Nutrition and of Home Economics Education	The staff of food service departments in hospitals, dietitians, etc.	Regional and local authorities; private firms
System is used by	The Administration; some departments of the universities of Uppsala and Stockholm	The Departments of Clinical Nutrition and of Home Economics Education	The staff of food service departments in hospitals, dietitians, etc.	Regional and local authorities; private firms
System is suited for	Nutrition surveys, especially large ones; diet planning; special diet planning; menu planning; research; education	Nutrition surveys; diet planning; special diet planning; menu planning; research; education	Nutrition surveys; diet planning; special diet planning; menu planning; research	Nutrition surveys; diet planning; special diet planning; menu planning; research; education
System is used for	Calculation of nutrients; calculation of statistical parameters; optimize recipes; food frequency; meal pattern	Calculation of nutrients	Calculation of nutrients; calculation of costs	Calculation of nutrients; calculation of costs; optimize recipes; nutrition labelling
Computer	Nord 100/CX	IBM 3033	ABC 800	FACII, ABC 80, ABC 800, IBM-PC
Program source	Own	SAS Gothenburg Computer Centre		
Program language	COBOL	SAS	BASIC	BASIC
Operating system	Sintran III VSE	GUTS terminal system, SAS	ABC DOS, EM DOS	Modified to each computer type
Data-base management system	MIMER	SAS	Own	
Access	Modem or terminals	Modem, terminals		
For sale:				
Nutrient data-base system	Yes; on magnetic tape	No	No	Yes; on floppy disk
Nutrient data base	Yes; on magnetic tape	No	No	No
Services	Nutrition surveys; menu planning; diet planning; calculation of nutrients, etc.			Calculations of nutrients, menu planning, etc.
Comments	A new version 1986, with e.g. 105 nutrients			

Table 3. Nutrient data banks in the Nordic countries: Iceland, 1985

System owner	The Icelandic Nutritional Council
Contact person	Jon Ottar Ragnarsson
Address	Sigtun 1
	IS-105 Reykjavik, Iceland
Telephone	+354 1 14 280
System name	
in operation since	1981
Nutrient data sources	Food composition tables – British, 1978, USA USDA *Agriculture Handbook No. 8-1;* own analyses
Number of (a) foods, (b) recipes, (c) aggregated foods, a-b, a-c	(a-c) 400
Nutrients/food	28
Other components/food	
Nutrition recommendation	
Recipe calculation	
Food yields	
Nutrient losses and gains; factors	
System is used by	The staff of the Icelandic Nutrition Council and the National Central Hospital
System is suited for	Nutrition surveys; planning of special diets; research
System is used for	Calculation of nutrients; optimize recipes
Computer	PDP 11/34
Program source	
Program language	FORTRAN 4
Operating data-base management system	RSX 11M, V.3.2 "Sequential file"
Access	
For sale:	
Nutrient data-base system	No
Nutrient data base	No
Services	
Comments	

Finland

Finland has at present five large nutrient data-base systems in operation, and an official one is being planned.

In the early 1970s a data-processing system for dietary surveys was established. This system has been used by the Department of Nutrition of the University of Helsinki, and by two research centres in Turku and Helsinki, mainly for research purposes. Since then the system of the university and of the research centres has developed into two different systems. The Departments of Nutrition and of Food Chemistry and Technology at the University of Helsinki, together with the National Public Health Institute, are now developing a new and more flexible data-processing system for food consumption surveys, while the research centres have designed their system for multi-purpose use.

The Research Centre for Household and Consumer Affairs and the State Computer Centre have programs for menu planning and nutrient calculation of diets.

Norway

At the end of the 1960s the Section for Dietary Research of the University of Oslo started to develop a nutrient data-base system. This is the oldest and most advanced system in Norway and a new version is now in use. In addition to this seven other systems exist. Five can be described as mostly educational: the two systems of the Tanum-Norli Publishing Co., the systems of the National Society for Nutrition and Health and the Norwegian College of Nutrition and Institutional Catering. The remaining two are mainly designed for diet and menu planning and analysis and are used preferably by nutritionists and dietitians.

The first steps towards an official nutrient data-base system have been taken and this will be ready for use in 1986.

Iceland

The Icelandic Nutrition Council owns the only nutrient data-base system in Iceland. This system cannot be considered official until it has been further improved.

Tables 2–5 give data on the national or main nutrient data-base systems in Finland, Norway, Sweden, and Iceland. For further information on Nordic systems see the *NORFOODS Directory of Nordic Nutrient Data Banks* [7].

REFERENCES

1. E. Abramson and B.-M. Andersson, *Kosttabell*, 7th ed. (Fesselte Studium, Stockholm, 1982).
2. P. Helms, *Fodevaretabeller* (Akademisk Forlag, Copenhagen, 1973–1975).
3. P. Helms, *Naeringstoftabeller* (Laegeforeningens Forlag, Copenhagen, 1978).
4. P. Koivistoinen, ed., *Mineral Element Composition of Finnish Foods: N, K, Ca, Mg, P, S, Fe, Cu, Mn, Zn, Mo, Co, Ni, Cr, F, Se, Si, Rb, Al, B, Br, Hg, As, Cd, Pb and Ash*, (*Acta Agriculturae Scandinavica*, Suppl. 22 (Stockholm, 1980).
5. Matvaretabell, *Statens ernaeringsrad*, 5th rev. ed. (Landsforeningen for Kosthold og Helse, Oslo, 1984).
6. A. Møller, comp., *Levnedsmiddeltabeller*, 2nd rev. ed. (Levnedsmiddelstyrelsen, Soborg, 1986).

7. *NORFOODS Directory of Nordic Nutrient Data Banks* (in press).
8. J. O. Ragnarsson, O. Reykdal, G. Thorkelsson, D. Johannsdottir, G. Sigurthorsson, E. Hilmarsdottir, and A. Sigurdsson, *Nutritional Value of Icelandic Processed Meat Products,* RALA Report, no. 107 (Utgefandi, Rannsoknastofnun Landbunadarins, Agricultural Research Institute, Keldnaholt, Reykjavik, 1984).
9. Statens Livsmedelsverk, *Livsmedelstabeller: energi och vissa naringsamnenskolupplaga,* 2nd rev. ed. (Statens Livsmedelsverk, Uppsala, 1984).
10. Statens Livsmedelsverk, *Livsmedelstabeller: energi och vissa naringsamnen,* 2nd rev. ed. (Statens Livsmedelsverk, Uppsala, in press).
11. O. Turpeinen, *Ruoka-ainetaulukko* [Food Composition Table] (Otava, Helsinki, 1985).
12. P. Varo, *Kivennaisainetaulukko* [The Mineral Element Table] (Otava, Helsinki, 1981).
13. P. Varo, R. Laine, K. Veijhalainen, A. Espo, A. Wetterhoff, and P. Koivistoinen, "Dietary Fiber and Available Carbohydrates in Finnish Vegetables and Fruits," *J. Agr. Sci. Finland,* 56: 49–59 (1984).
14. P. Varo, R. Laine, K. Veijalainen, K. Pero, and P. Koivistoinen, "Dietary Fiber and Available Carbohydrates in Finnish Cereal Products," *J. Agr. Sci. Finland,* 56: 39–48 (1984).

17
Food Data in Canada:
The Canadian Nutrient File

PAMELA C. VERDIER

Bureau of Nutritional Sciences, Food Directorate, Health Protection Branch,
Health and Welfare Canada, Ottawa, Ontario, Canada

INTRODUCTION

The first *Table of Food Values Recommended for Canadian Use* was produced by the Department of Pensions and National Health in 1944 [6]. The second edition [3] contained the statements, "There is scarcely one single food for which complete analyses made in a Canadian laboratory are available" and "Where Canadian values were available in sufficient range they have been used." This edition used data from USDA *Agriculture Handbook No. 8*, 1950 [8]. In the late 1960s, the information was revised and reformated into a computerized version for calculating the 24 hour-recall food consumption portion of the Nutrition Canada Survey, 1971–1972 [1]. The source of the data was, in most cases, USDA *Handbook No. 8*, 1963 [8]. "Canadian only" foods were entered, examples being all baked goods sold in the province of Newfoundland; these are made with "Newfoundland flour," a calcium-enriched product. Cooking-yield factors were applied to raw foods where cooked values were not available, some nutrient analyses were performed and appropriate levels of fortification of various foods changed to meet Canada's food and drug regulations. After the survey, this data base was made available to persons able to use it, and various other nutrient values (such as zinc and copper in a limited number of foods) were added over the years to meet the needs of the Bureau of Nutritional Sciences. By 1978, no more foods or nutrients could be added and it was decided to reformat the entire tape into a more flexible and accessible form, to be called the Canadian Nutrient File (CNF).

METHODS

In setting up the new file, a three-subfile layout was designed to facilitate easy access and manipulation of the data. The three subfiles were: the Food Name Subfile, the Nutrient Name Subfile, and the Nutrient Amount Subfile. USDA was then in the process of issuing updated *Handbook No. 8* food groups, and so, in reformating the

Table 1. Example of Food Name Subfile

Food code	Year added	Food description	Nutrition Canada code	Flag	Factor			Refuse (%)
					1	2	3	
012ª012	81	Cheese, cottage, creamed	0647	1ᵇ	*ᶜ	100 ML SM curd	100 ML LG curd	0

a. Food with nutrients added.
b. Improved data added after 1981.
c. Star indicates that 100 ML edible portion conversion factor is available for this food (in this case factor 1 is for a "not further specified" food description).

data for the new file, a coding system incorporating the new USDA codes and containing an extra digit for Canadian use was developed.

The Food Name Subfile (table 1), in English and French, includes the food name, the new food code, the old Nutrition Canada Survey food code based on *Handbook No. 8*, 1963 (the year the food was added to the data base) [1], an editorial flag, and four conversion factors. The extra digit was added to the new USDA food codes to indicate whether the food was the same as in *Handbook No. 8* (in which case the number was zero) or what modifications were made (foods with some nutrient values entered for Canadian levels of fortification, values analysed in Canada, calculated values, etc.). "Canadian only" foods were coded similarly and incorporate within the code the closest USDA food description. The Nutrition Canada Survey food codes, if applicable, were included for the convenience of those persons used to the Nutrition Canada Survey conversion factors. The year the food was added to the base is included to facilitate retrospective studies. When an update is needed because of a change in formulation of a food, the code is left intact and a new food record is created with the appropriate year. The editorial flag indicates that improved data have been added although the food itself has not changed. Portion sizes other than the "100-gram edible portions" are available by multiplying the 100-gram portion values by four factors. Multiplying by factor 1 gives the edible portion in 100 ml (for volumetric measurements); factors 2 and 3 have 15-digit descriptors for common portion sizes, such as: 1 large egg, 100 ml puréed, slice $10 \times 10 \times 0.2$, etc. (all dimensions are in centimetres and all factors have been recalculated for metric volumes); and factor 4 gives 1 kg as purchased.

The Nutrient Name Subfile (table 2), in English and French, contains a description of the nutrient, the three-digit USDA nutrient code, and the measure associated with each nutrient [7]. Also provided with the tape is a listing of the number and percentage of foods in the file that have values for each nutrient. This assists the user in deciding whether a meaningful survey of a specific nutrient can be undertaken.

The Nutrient Amount Subfile (table 3) identifies the food by the new Canadian food code and gives the nutrient codes with all available nutrient data expressed per 100-gram edible portion. A single-digit flag is added to each nutrient value to show how the item differs from the USDA value because of added, calculated, substituted, or imputed nutrients. When there is no flag, no change has been made.

Canada still has little original data on Canadian foods, although some Canadian research data have been incorporated into the *Handbook No. 8* series (both 1963

Table 2. Nutrient Name Subfile examples

No. of foods	Percentage	Nutrient code	Unit	Nutrient name
3,060	99.8	0203	g	Protein
3,026	98.7	0206	g	Fibre, crude
146	4.8	0273	g	Fibre, neutral detergent
767	25.0	0806	g	Fibre, total dietary (calculated)[a]
1,672	54.5	0309	mg	Zinc
172	5.6	0339	μg	Vitamin D[b]

a. Values calculated from Paul and Southgate [7] by the Ludwig Cancer Institute, Toronto, Ontario.
b. Vitamin D in μg to conform with Canadian recommendations.

Table 3. Nutrient Amount Subfile examples

Group no.	Item no.	Nutrient code	Mean	Standard error	No. of observations	Flag
01	2012	0203	12.49	0.1592	69	
		0312	0.02			5[a]

a. 5 = nutrient imputed from a similar food.

and present). Folic acid values in many foods and pantothenic acid in cereals are examples. As each new food-group book is issued by USDA, the data are examined for relevance to the Canadian situation by consultation with experts from Agriculture Canada, university departments, and the appropriate food manufacturers and marketing boards. Canadian research values are entered, where available, and levels of fortification are changed to meet Canadian food regulations. Breakfast foods in Canada are not as highly fortified as in the United States, and, consequently, Canadian manufacturers' values, supplemented by Canadian research data, are used for this food group. Certain *Handbook No. 8* values are, nevertheless, added when the determinations are not available from a Canadian source and the food descriptions are the same as in the handbook (e.g. copper and magnesium for Kellogg's Corn Flakes). Some changes are made to meet the Canadian food situation: canola (rape-seed) oil is used in all commercial salad dressings and mayonnaises in Canada, and Canadian analytical values for fatty acids are therefore entered for these products. Some foods with completely calculated nutrient values are included: 2 per cent evaporated milk (calculated from USDA *Handbook No. 8* values for evaporated milk, whole, and skim). Imputations of nutrients from a similar food (such as zinc and copper) are kept to a minimum and no attempt is made to impute "missing values" for foods that have incomplete data. All Canadian entries, substitutions, calculations, or imputations are flagged and readily identifiable. Food names are changed to reflect Canadian usage: "Cheese, processed, American" becomes "Cheese, processed, Cheddar." "Canadian only" foods are added where data are available and foods not on the Canadian market (such as unenriched flour and flour products) are deleted.

The new format includes metric portions only. As Canada is a bilingual country, the users' manual accompanying the nine-track, 1600 BPI tape is in English and French, the two official languages. Instructions are also provided for the computer programmer and include a printout of the Food Name and Nutrient Name Subfiles. The tape, written in COBOL, may be accessed by any computer language but no software program is provided. The file is created on an Andahl 470-V8 computer and the tapes, which are updated annually from the USDA data tapes, are for sale. The first tape was issued in December 1981.

In the early 1950s, a table of nutrient values for 185 foods was prepared as part of a nutrition education booklet, *Healthful Eating* [2], which was written for teachers, nurses, and the general public. Because of the large demand for this publication, an expanded separate leaflet called *Nutrient Value of Some Common Foods* [4] was issued. This has continued to expand and is available, free, in both official languages. It now provides values in customary household measures for moisture, energy, and 17 nutrients. A bilingual tape, containing values for the 608 foods and the nutrients included in the present publication, was prepared for persons wanting a smaller, more manageable data base. It can be sorted either alphabetically or by food group and is based on 100-gram edible portions, with one factor providing the common household measure. It is also to be updated annually.

DISCUSSION

The reformated file was designed to provide all reliable nutrient data available for the described foods and to include features that would facilitate its use. It was developed for Health and Welfare Canada to calculate nutrient intakes in food consumption surveys and to analyse menus for food-fortification or food-substitution studies. Agriculture Canada also had a need for such data and it was hoped that it would provide a national file on which all Canadian studies could be based. User comments were solicited from the beginning and changes have been made in the format to facilitate use. A major change was from seven conversion factors, described in the user's manual accompanying the tape (in the first two editions), to the four factors described on the present tape.

A National CNF User's Workshop was organized in June 1984 in Toronto, Ontario, to promote the general use of the file and share experiences in using it. Development of software programs requires the skills of both a nutritionist and a computer programmer. Agriculture Canada, with nutritionists, statisticians, and programmers, acquired the tape and has developed a very extensive programme: their nutrient-analysis computer program (AGNAP) has both nutritional and statistical input. Most of the nutritional data in AGNAP come from the CNF, but there is also a cooking-yield file for "in-house" research information. Six types of reports can be produced: recipe/menu analysis, food ranking by amount of specific nutrient, cost of food per unit of nutrient, apparent nutrient intakes, nutritive value of food purchased, and nutritious food-basket data manipulations. Statistics Canada provides data for four other reference files: a population file, providing estimates of demographic groups within the Canadian population; a food disappearance data file, outlining the apparent per capita food consumption in Canada between 1960 and 1983 (updated annually); a family food expenditure data file, including the quantities of food purchased by various socio-economic groups within Canada (derived from a continuing market basket survey); and a retail food prices file, based on Statistics Canada's consumer price index. From these are derived three programs: the Apparent Nutrient Intakes program, the Nutritive Value of Food Purchased program, and the Agriculture Canada Nutritious Food Basket development program. At the other end of the scale, a nutritionist in a rural community, with specific needs for the evaluation of nutrient intakes in hospital food services, nursing homes, and boarding-houses, wrote a program on a microcomputer including nutrient assessment of individuals and menu planning, and now has extended use with a modem connection to a local university to access the entire CNF.

Several users have had special programs written for such projects as a regional hospital food-services group that supplies prepared frozen foods daily to seven hospitals with a completely computerized program for food purchase, quality control, menu planning, ingredient delivery, assembly, cooking, freezing, and storage prior to delivery. Attached to these programs are nutrient compositions of all foods as delivered. A food-service organization, providing bulk foods to boarding-houses and nursing homes, includes computerized recipes, menus and probable nutrient intakes.

Most users buy a software program, including all or part of the CNF. Services are also available for the submission of data to a university computer centre for computation. As the capacity of microcomputers expands, more and more programs are able to store and access the entire file. Programs are becoming more sophisticated and various user needs are being met. Canadian programmers, using their experience in developing packages for the CNF, are working on software systems for third-world countries.

A further extension of the CNF is in progress with the development of a data file for use in dietetics in Canadian hospitals. Louise Bell, a nutritionist at the Hospital for Sick Children in Toronto, is modifying the file to include imputations of missing values, standard foods, brand-name foods including infant formula and enteral feedings, special dietetic products, and vitamin and mineral supplements, with unused foods being deleted. The first issue will have complete data for all foods covering 24 nutrients.

Interest is developing in identifying areas where Canadian data are needed for differences in the foods, and in defining research programmes to fill these gaps. Contracts have been made out by Agriculture Canada for meat [5] and other analyses, and an ongoing assessment of the Canadian food supply is being conducted by Health and Welfare Canada, in conjunction with the Total Diet Study, regarding specific nutrients where differences from the US data might be expected or where data are now missing (e.g. fat levels in meats and neutral detergent fibre in many foods). The Field Operations Directorate of Health and Welfare Canada already samples and analyses certain selected foods and nutrients each year and provides data that are useful in checking the applicability of the USDA data. Over the years, such research will help to increase the proportion of Canadian data in the CNF.

Data are missing for northern animal, fish, and plant foods. Recommendations to resume consumption of available "land foods" in the north must be supported by data on nutrient content. As only a small part of the population consumes such foods, little research interest exists, and an exchange of data with other northern countries is being sought.

The CNF format is compatible with that of USDA and will form part of the NOA-FOODS regional network.

REFERENCES

1. Department of National Health and Welfare, *Food Consumption Patterns Report* (Bureau of Nutritional Sciences, Health Protection Branch, Ottawa, 1976).
2. Department of National Health and Welfare, *Healthful Eating* (Nutrition Division, Ottawa, 1952).
3. Department of National Health and Welfare, *Table of Food Values Recommended for Use in Canada* (Nutrition Division, Ottawa, 1951).
4. Health and Welfare Canada, *Nutrient Value of Some Common Foods* (Health Services and Promotion Branch and Health Protection Branch, Ottawa, 1983).
5. S. D. M. Jones, "Chemical Composition of Selected Cooked Beef Steaks and Roasts," *J. Can. Diet. Assoc.*, 46: 40–44 (1985).
6. M. E. Macbeth and L. B. Pett, *Table of Food Values Recommended for Canadian Use* (Nutrition Services, Department of Pensions and National Health, Ottawa, 1944).
7. A. A. Paul and D.A.T. Southgate, *McCance and Widdowson's: The Composition of Foods*, 4th ed. (HMSO, London, 1978).
8. US Department of Agriculture, "Composition of Foods: Raw, Processed, Prepared," *Agriculture Handbook No. 8* (Science and Education Administration, USDA, Washington, D.C., 1950, 1963).

Other Considerations

18
A System for Evaluating the Quality of Published Nutrient Data: Selenium, a Test Case

JOANNE M. HOLDEN, ANITA SCHUBERT, WAYNE R. WOLF, and GARY R. BEECHER

Nutrient Composition Laboratory, Beltsville Human Nutrition Research Center, Beltsville, Maryland, USA

INTRODUCTION

Food composition data are used by nutritionists, dietitians, and epidemiologists to assess the adequacy of the diets of population groups, subgroups, and individuals. They are used to determine federal and state government policy regarding food and nutrition programmes and other public health efforts. Data relative to the assessment of nutrients and contaminants such as pesticides are used to formulate the policies of regulatory agencies such as the Food and Drug Administration (FDA). In addition, the food industry uses food composition data to assist in food labelling, quality assurance, and product development. Ideally, data to be used for such varied and far-reaching assessment and policy formulation should be accurate and precise by analytical standards, and should represent both the qualitative and quantitative standards and the qualitative and quantitative distribution of nutrients or other components found in the food supply consumed by the group or individuals to be studied. Because data obtained in such a manner that they meet all of these requirements are scarce, critical evaluations of the limitations of available published data are needed.

Food composition data of variable quality and quantity can be collected from many different sources. They can be obtained from manufacturers of food products; these may include individual values for replicates or mean values for specific samples. Data can be taken from product labels. Certain data are obtained directly from the analysis of foods purchased and sampled specifically for the purpose of determining their composition. These studies may or may not be published. Food composition data may also be gathered from the scientific literature as the indirect results of the development of analytical methodology, animal feeding trials, soil treatment trials, and bioavailability studies. While these published studies may accomplish stated objectives, not all the data may be suitable for use in food composition tables and data banks. Since the requirements of data users can be diverse, the suitability of specific data for inclusion in a data base should be evaluated according to objective criteria that are known to users of the data. Some indicator of data quality is necessary for each nutrient in each food to provide the data user with a measure of the reliability and usefulness of specific values in the data base.

The main objective of this work was to develop a set of criteria that could be applied to published analytical data (a subset of available nutrient data) for any specific nutrient or component in foods. Selenium (Se) was selected as a test case for these criteria because of the current high level of interest in this nutrient and the need for accurate assessment of Se intake in a number of ongoing human studies. Furthermore, published Se data have been generated by a limited number of satisfactory analytical methods and constitute a finite data set suitable for testing such a system of criteria.

A secondary objective was to provide analysts with a set of guidelines for designing nutrient composition studies and reporting their results. Such guidelines can also be used by journal editors and their reviewers to evaluate scientific papers submitted for publication and to elicit details of a study important to compilers of nutrient composition data.

Criteria developed specifically for Se were applied to the body of Se data available in the scientific literature. Ratings determined in the process were combined to yield a confidence code (CC) for the mean Se value of each food item reviewed. This paper will describe these criteria and their application, and give several examples. (A table of foods with their Se concentrations, respective CCs, and specific references will be published separately.)

BACKGROUND

In 1980 the need for objective criteria for evaluating food composition data was recognized by USDA workers at the Nutrient Composition Laboratory of the Agricultural Research Service and the Nutrient Data Research Branch of the Human Nutrition Information Service. Discussions led to the development of data quality criteria which were used to evaluate iron data for publication of the provisional table *Iron Content of Food* [8].

The various studies for specific foods were evaluated according to criteria in three categories: (a) documentation of analytical method, (b) sample handling and appropriateness of analytical method, and (c) (analytical) quality control. Scores for these criteria led to the assignment of a CC, which appeared in the iron table adjacent to the mean iron concentration. For the first time, users of a nutrient composition table were provided with a measure or degree of confidence in the mean value for that particular food. Asterisks attached to a CC indicated either a limited number of sources or the extent of variability.

Recently, Stewart [26] reiterated the importance of evaluating nutrient data for inclusion in data bases. He recommended several critical activities that contribute to the generation of high-quality analytical data on foods, including (a) development of appropriate validated analytical methods, (b) use of sound food-sampling techniques to ensure representativeness of samples, and (c) use of appropriate quality-control systems in conjunction with validated methodologies to ensure the production of validated composition data.

Although we know of no other similar effort in the field of food composition data, some researchers in other fields have been concerned with systems for the evaluation of data quality. W. Mabey et al. [12] have recently published a similar system of

178

criteria to be used for the evaluation of quantitative data to be included in a data base for environmental fate assessment. For each of four major categories, specific check-lists of criteria were presented to permit a thorough evaluation of each datum, resulting in the derivation of a data quality index. This index indicates to the data-base user that an objective evaluation of data quality has been done and should help to educate users and data generators and improve the overall excellence of the data.

PROCEDURE

The system of objective criteria described here for the evaluation of published nutrient composition data reflects the basic concepts described by Exler [8] and incorporates

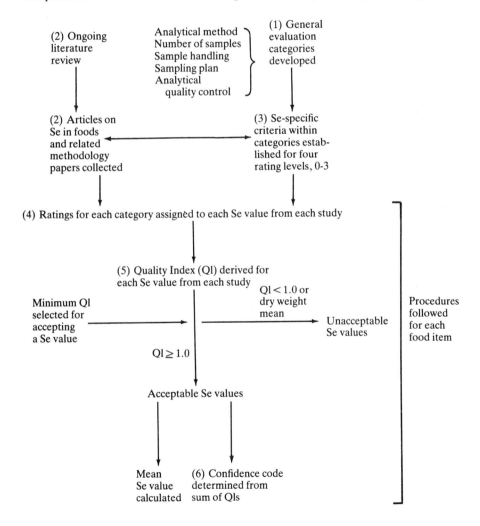

Fig. 1. Flowchart of evaluation system for published selenium data.

the critical aspects cited by Stewart [26]. The system involves several stages (fig. 1): (a) development of categories of criteria to include key issues relative to analytical methodology and sampling; (b) review of literature and collection of those papers that reported the Se content of foods and related methodology; (c) definition of criteria in each category at four rating levels to reflect specific concerns that pertain to the evaluation of literature reporting the Se content of foods; (d) assignment of ratings for each criterion by study and by food item; (e) derivation of the quality index for each study; and (f) derivation of mean value and CC for the Se content in each food by combining data from the acceptable studies.

Five general categories were developed to evaluate the data for each food item: (a) number of samples, (b) analytical method, (c) sampling handling, (d) sampling plan, and (e) analytical quality control. These categories identify the issues essential to any study related to the composition of foods. A rating scale of 0 (unacceptable) to 3 (most desirable) was established. Within each category, the criteria for each rating level were defined specifically for Se. Establishing these criteria demanded knowledge of accepted methodology, sample handling procedures, and quality-control measures specific to this nutrient. In addition, knowledge of statistical methods was required for the sampling plan and number of samples categories. The rating criteria are outlined in table 1 and described in detail below. In general, within each category the level of documentation and appropriateness of procedures are addressed.

An extensive literature search yielded approximately 65 papers (from 33 different journals, reports, proceedings, and books) published after 1960 that report original analytical Se food data. Several references include data from more than one study. Papers published prior to 1960 were collected for historical purposes, but the data were not included in this evaluation because of difficulty in assessing the validity of the methodology used and conceivable lack of relevance of those data to current studies of food consumption due to possible changes in the food product during the last 25 years. The methodology papers referenced in the data articles were also collected. In addition, data from recent FDA Total Diet Study analyses, unpublished at the time of this evaluation [23], were included because this programme is one of the few that has analysed food as eaten, i.e. cooked foods and mixtures. Data from fiscal years 1982/83 and 1983/84 were treated separately, providing two sets of means for each food item analysed. Since the focus of this work was the Se content of foods frequently consumed by Americans, only those studies that analysed foods grown, processed, or sold in the United States and Canada were collected.

As previously mentioned, the data from each study were rated on the criteria on a scale of 0 to 3. In general, 0 was assigned when information was not adequate to permit evaluation of data for use in food composition data bases, or when certain procedures or practices were inappropriate. A 3 was assigned when procedures were well-documented and appropriately applied.

CRITERIA

Number of Samples

Statistical rigour requires that the appropriate number of samples for a study be a function of the nutrient variability within the population of each food item [6]. Both

Table 1. Data quality criteria[a]

Ratings	3	2	1	0
Number of samples	>10; SD, SE, or raw data reported	3–10	1–2; explicitly stated or not specified	—
Analytical method	Official fluorometric (ref. given) or other method documented by a complete published write-up with validation studies for foods analysed, including use of appropriate SRM where available, 95–105 per cent recoveries on food similar to sample analysed in same or other paper; Se concentration above quantitation limit of the method	Modified fluorometric or other method, some documentation, incomplete validation studies for foods analysed; must include 90–110 per cent recoveries on food similar to sample analysed (or good recovery but no statistics given), and/or use of other method (official fluorometric, isotope dilution or NAA) on same sample with good agreement (within 10 per cent)	Non-fluorometric method, partially described; 80–90 per cent or >110 per cent recoveries on food similar to sample; or use of comparison method or recoveries on food only somewhat related to sample (animal/plant)	No documentation of method, no ref. or inaccessible ref. given, no validation studies, or poor agreement (>10 per cent) of test method with comparison method on same sample
Sample handling	Complete documentation of procedures, including analysis of edible portion only, validation of homogenization method, details of food preparation, and storage and moisture changes monitored	Pertinent procedures documented, including analysis of edible portion only; procedures seem reasonable but some details not reported	Limited description of procedures, including evidence of analysis of edible portion only	Totally inappropriate procedures or no documentation of criteria pertinent to food analysed
Sampling plan	Multiple geographical sampling with complete description; sample is representative of brands/varieties commonly consumed or commercially used	1 or 2 geographic areas sampled; sample is representative	Sample representative of small percentage of US and/or origin not clear	Not described or sample not representative
Analytical quality control	Optimum accuracy and precision of method monitored and indicated explicitly by data	Documentation of assessment of both accuracy and precision of method; acceptable accuracy and precision	Some description of minimally acceptable accuracy and precision of method	No documentation of accuracy and/or precision

a. See text for complete description of criteria.

181

intrinsic and extrinsic sources of variation affect the levels of Se in foods. However, when evaluating published data, one rarely has access to the magnitude of specific sources of variation for a given food. The standard deviation, when given, is an indication of total variance. Some variability is assessed indirectly by the other categories of criteria and includes: systematic error intrinsic to the analytical methodology or sample handling, variability attributable to differences in brands and varieties of foods analysed, and errors in analytical accuracy and precision in the execution of the analytical method. A statistical formula can be used to estimate appropriate sample size given a mean, standard deviation, and level of acceptable error [6]. However, the error term selected would depend upon the concentration of the nutrient in a given food item and the detection and quantitation limits [11] for the analytical method. Such a judgement could be made given sufficient data for each food. Using a predetermined coefficient of variation of 20 per cent as the limit of acceptable error, we estimated appropriate sample size for a limited number of studies where standard deviations were reported. Our estimate and the actual number of samples analysed in each study were comparable.

In the absence of adequate information for many studies, particularly standard deviation data, we chose to make a somewhat subjective judgement on the sample size limits for each rating: a rating of 1 for one to two samples or when number of samples is not specified; a rating of 2 for three to ten samples; and a rating of 3 for greater than ten samples and inclusion of the standard deviation, standard error, or raw data from which a standard deviation can be calculated. A rating of 0 is not applicable in this category. As documentation improves, it will be possible to evaluate the appropriateness of the number of samples analysed based on statistical considerations.

Analytical Method

Several issues related to the method of analysis are important with regard to ratings on analytical method. Documentation of whatever method is used is primary: suitability of method cannot be determined if no description or reference to details in another paper is included. A rating of 0 is assigned where the method is not described, no reference is given, or a reference is generally unavailable. In some cases the use of "official" methods for the analysis of the specific foods merits a higher score. The use of the official fluorometric method for Se analysis, as published in the handbook of the Association of Official Analytical Chemists (AOAC) [27], is rated a minimum of 2. However, use of an "official" method does not preclude attention to a second important issue: validation of the method.

Validation of the test method for the general matrix (e.g. meat, grain, fat), and preferably for the specific food item in question, is necessary to show that accurate results can be obtained. Use of recovery trials on the same or a similar food is one aspect of method validation. Higher ratings are earned for recovery close to 100 per cent and similarity to the food analysed of the food on which the recovery trials are done. The use of a highly regarded second analytical method on the same or a similar food is another aspect of method validation. In the case of Se, highly regarded analytical methods are the fluorometric method approved by the AOAC, neutron activation analysis, and isotope dilution-mass spectrometry. The analytical results of this second method must be in good agreement with the results of the test method, or a 0 is assigned. Good agreement is defined according to our adaptation of Stewart's general recommendation that the values obtained by comparable methods should be within 10 per cent of each other if a daily intake of the food provides greater than 5 per cent of the US RDA for that nutrient [25]. Since there is no US RDA for Se, we have used 5 per cent of the lower end of the estimated safe and adequate daily dietary intake of 50–200 μg of Se, as recommended by the Food and Nutrition Board of the National Research Council [7].

The analysis of a standard reference material (SRM) by the method in question and comparison of the value obtained with the certified value and the range of estimated uncertainty of the SRM is useful for validation when the matrix and Se concentration of the SRM are similar to that of the food in question [28]. However, the small number of SRMs that are certified for Se has confined this aspect to a rating of 3 for analytical method. As the availability of SRMs with a variety of matrices increases, this aspect of the criterion will be required for ratings below 3.

Finally, evidence that the analysis is carried out above the quantitation limit of the method, as defined by the American Chemical Society Committee on Environmental Improvement [11], is required to assure that the method can determine expected levels of Se in the food to be analysed. For a rating of 3, the quantitation limit must be defined and be below the Se level reported for the food in question.

Other issues with regard to analytical method are of some concern, but were considered secondary to the main points. One such issue is that of the size of the samples that are analysed. Sample size must be adequate relative to the sensitivity of the method so that (a) the concentration is above the quantitation limit, and (b) the sample analysed is representative of the whole food item. This concern was not included in analytical method requirements.

In summary, three concerns must be satisfied for data to be rated 3 in analytical method: (a) a complete description of the method in the same or another accessible publication; (b) validation studies for the food in question, which can consist of either recovery trials with 95 per cent to 105 per cent recovery of Se or comparable results with use of a second method that is highly regarded, as well as use of an appropriate SRM when available; and (c) reported analytical values above the defined quantitation limit of the method.

Sample Handling

How a sample is handled from the time of acquisition to the time of analysis is critical for general nutrient stability. For example, preventing the loss of volatile components is important for maintaining relative nutrient concentration, a factor of importance for Se. Therefore, documentation of sample handling protocol is essential to evaluate data pertaining to the nutrient composition of a food. Lack of documentation of sample handling procedures or use of inappropriate procedures is rated 0. Se contamination of food samples via utensils, cooking ware, grinder, or containers is not a problem, in contrast to the analysis of other inorganic nutrients such as zinc and chromium. However, details of homogenization, temperature control, and other aspects of sample preparation must be known to evaluate the representativeness of a sample aliquot taken from a large batch of prepared material.

Analysis of the edible portion must be reported for a rating of 1 or higher. For example, some canned foods must be drained, raw fruits and vegetables must be peeled or cored, and meats must be boned and trimmed of fat if these foods are generally eaten that way. Thorough homogenization of the food is critical for food items with diverse constituents. Examples of such foods are: breaded and fried fish or poultry, food mixtures, and fruits or vegetables eaten with skin or seeds. Ideally, thoroughness of homogenization is checked by analysing portions from various parts of the final mixture. Additional factors that should be reported for the highest rating are: detailed description of the food, including processing methods (e.g. whether rice was polished, unenriched, or instant); cooking method (if any); general storage conditions, e.g. frozen foods kept frozen, fresh foods analysed soon after pick-up; and measurements of moisture/volatiles content.

Sampling Plan

The sampling plan of the study reflects the representativeness of the samples with regard to the brand or cultivar, method of preparation, and geographic origin of the food. Is the particular food item typical of what many Americans eat? No description of the sampling plan or the use of a non-representative sample is rated 0. This would include a food grown under experimental soil conditions, food grown in someone's home garden, or food prepared in an unusual way. Data from Canadian studies were evaluated because some foods sold in Canada are grown in the United States – e.g. fruits and some grains – and therefore can be representative of what Americans eat. Data for foods grown on Canadian soil but not exported to the United States were given a lower score (0 or 1) for sampling plan since the concern of this work was foods consumed by Americans. The use of popular brands and frequently consumed forms of foods was rated 2. Obtaining representative samples from supermarkets in more than two well-populated areas was rated 3. For fresh foods obtained from growers or producers, representativeness of cultivar and geographic source was assessed by referring to *Agricultural Statistics* [29].

Analytical Quality Control

Information that details acceptable accuracy and precision in the day-to-day execution of an analytical method is necessary for evaluating the quality of nutrient data. Accuracy and precision are each rated as separate aspects of analytical quality control. For each datum, the lower rating of the two aspects determines the rating in this category.

Accuracy is the degree to which an analysed value represents or estimates the "true" value. An investigator must demonstrate that the method is capable of accurately determining the nutrient level in a particular food item; that is, a method must be validated for each general matrix, as described in the analytical method section above. Once the method has been validated, it must be carried out appropriately each time an analysis is performed. Accuracy in the day-to-day use of a method is one of the two elements that must be monitored and reported for a study's data to be rated favourably in analytical quality control.

Day-to-day accuracy is monitored by analysis of a quality-control material that is similar in matrix and nutrient concentration to the test sample. Analysis of such a material should be included with each batch of unknowns or on each day of analysis if several batches are run in a day. Quality-control materials can be SRMs such as those available from the National Bureau of Standards (NBS), which are certified for specific nutrients, or they can be secondary reference materials, i.e. materials

Table 2. Accuracy requirements for secondary reference materials

Analysed mean value of secondary reference material should fall within the reference mean value ±	To receive a rating of
2 standard deviations	3
2½ standard deviations	2
3 standard deviations	1

developed especially for a study and characterized by one or more methods, including reference methods. NBS SRMs currently available and certified for use in evaluating the accuracy of a method for the determination of Se in foods are: Orchard Leaves 1571 [17], Wheat Flour 1567 [20], Rice Flour 1568 [19], Bovine Liver 1577a [15], Oyster Tissue 1566 [18], and Non-Fat Milk Powder 1549 [16].

When an NBS SRM is used to judge accuracy, the analysed mean value must fall within the mean plus or minus the estimated uncertainty, as stated in the NBS certificate, for the study to be rated a 3. Use of an SRM with results falling outside the range of estimated uncertainty is rated 0. This apparently stringent rule is based on the sources of SRM values. Many certificates of analysis do not label the uncertainty in statistical terms such as confidence intervals or a certain number of standard deviations. For example, the Rice Flour (no. 1568) certificate states [19]:

The estimated uncertainty is based on judgement and represents an evaluation of the combined effects of method imprecision, possible systematic errors among methods, and material variability for samples 400 mg or more. (No attempt was made to derive exact statistical measures of imprecision because several methods were involved in the determination of the constituents.)

When a secondary quality-control material is used, a reference mean and standard deviation should be obtained by analysis of the material by the same or another laboratory using a reference method. To monitor a method's accuracy, the formulated reference material should be analysed with each batch of unknowns or on each day of analysis, comparing the results to the reference mean and standard deviation as shown in table 2.

The review of Se references carried out for this work revealed that a reference material was frequently analysed at the outset of a study, serving to validate the method before analysis of unknowns begins. Occasionally a mean and standard deviation for a quality-control material were reported, but usually documentation was not sufficient to determine whether the material was analysed with each batch of unknowns. In such cases it was assumed that the investigator used the reference material only to validate the test method rather than to measure day-to-day accuracy. In this instance, an accuracy rating of no higher than 1 was assigned.

The other half of analytical quality control is the level of precision; the aspect of concern here is the amount of variability about the mean value associated with the day-to-day execution of a particular method. The indication of day-to-day variability can be determined only when the analytical method is monitored continuously through the use of a quality-control material similar in matrix to the unknowns to be analysed. Like accuracy, the day-to-day precision of a method is matrix-dependent.

Precision is usually measured by calculating the per cent coefficient of variation (% CV), also known as per cent relative standard deviation (% RSD), from the

Table 3. Assignment and meaning of confidence codes

Sum of quality indices	Confidence code	Meaning of confidence code
>6.0	a	The user can have considerable confidence in this value.
2.4 to 6.0	b	The user can have confidence in this value; however, some problems exist regarding the data on which the value is based.
1.0 to <3.4	c	The user can have less confidence in this value due to limited quantity and/or quality of data.

mean and standard deviation (SD) of several replicates of a sample: % CV or % RSD = SD divided by mean × 100 per cent. The lower the % CV, the more precise the analysis. Our limits for rating % CV are: 5 per cent or less for a 3, 10 per cent or less for a 2, and 15 per cent or less for a 1. The % CV, calculated from replicates analysed within a given laboratory, includes variability attributable to instrument and technician performance and to the method. The precision of a method could be poor in the hands of one investigator and acceptable in the hands of another.

When the ratings for accuracy and precision are the same, that rating becomes the analytical quality-control rating. When accuracy and precision ratings differ, the lower rating is assigned for the overall analytical quality-control rating for the study. Data from a study with incomplete documentation are rated no higher than 1 in this category. Lack of any documentation or unacceptable precision or accuracy earns a 0 in analytical quality control.

CALCULATION OF THE MEAN SE VALUE AND CONFIDENCE CODE

The quality index (QI), a measure of the overall quality of data from a single study, was derived in one of two ways: (a) when analytical method is rated 0 or when three or more categories are rated 0, the QI for that study becomes 0; or (b) when those conditions do not exist, the ratings for each of the five categories are averaged. Thus, the QI can range from 0 to a maximum of 3.0.

The mean Se values reported on a fresh weight basis from those studies which have a QI equal to or greater than 1.0 are averaged together to obtain a mean Se value. Values reported on a dry weight basis cannot be combined with values reported on a fresh weight basis, i.e. foods as eaten, unless moisture levels are included with the dry values to permit calculation back to fresh weight. If no moisture levels are reported, dry values are excluded regardless of their QI.

A confidence code (CC), assigned to the mean Se value for each food, indicates the degree of confidence a user can have in the mean value. It is determined by summing the QIs equal to or greater than 1.0 among the various studies evaluated for a given food item, and then referring to table 3 for the corresponding CC. The basis for the CC is the necessity of confirming the results of one report by other investigators in order to be considered valid. Thus, data from a minimum of three studies with a sum of QIs of 6.2 are required for a mean value to be assigned a CC of *a*. The cut-off point between *b* and *c* of 3.4 was made by dividing approximately equally all the possible sums of QIs, i.e. from 1.0 to 6.0.

RESULTS

Table 4 provides the Se data which were collected and evaluated for three foods: white rolls, whole milk, and canned crab. These foods were selected to show varying levels of Se among foods and within foods, different combinations of ratings, the typical range of QIs, and an example for each CC.

Analyses of white rolls were reported in three references. In reference 22, analyses were carried out on samples of the same product from two different sets of cities. Zeros were assigned in two categories: in analytical quality control because of poor precision in the second set of collections in reference 23, and because of lack of any documentation in references 3 and 22; and in sampling plan because of the Canadian origin of the product in reference 3. One-half of the ratings were 2, while 1 and 3 were assigned less frequently. The data from all four studies (three references) were found to be acceptable, and the four Se values were used to derive the mean value of 34 µg/100 g, with minimum and maximum values of 21 and 61. The CC of *a* based on the sum of the QIs, 6.4, indicates that one can use this value with considerable confidence.

Sample analyses of whole milk were presented in 11 references which contained 15 studies. However, data from 11 studies were judged unacceptable based on a 0 rating in analytical method: lack of documentation, lack of validation of the method for milk, or reported Se levels below the stated quantitation limit. Out of the four remaining studies [2, 3, 4, 13], data from two [2, 3] were rated 0 in sampling plan due to their Canadian origin, while all were rated 0 in analytical quality control: references 3 and 4 reported no quality-control measures, while references 2 and 13 did not address both issues of accuracy and precision. Reference 4 does not explicitly report the number of whole milk samples analysed for Se; only the total number of samples is stated (103) which includes three types of milk in addition to whole milk. Although it is likely that a large number per type was analysed, the number of samples category received a 1 due to insufficient documentation. In two studies [2, 3], only one sample (in duplicate) was analysed. The mean Se level in whole milk, based on the data from four acceptable studies, is 1.6 µg Se/100 g, with a minimum and maximum of 1.1 and 2.5. The sum of QIs is 5.6, which establishes a CC of *b*, indicating that a user of this datum can have some confidence in the value.

Canned crab samples were analysed by two sets of workers, with a Se value from only one study [5] receiving an acceptable QI. The investigators reported no analytical quality-control measures (rated 0), analysed duplicate portions of only one sample (rated 1 in number of samples), and recovered only 80–90 per cent of Se in recovery trials (minimal validation results rated 1 in analytical method). Thus, the concentration of 22 µg Se/100 g is based on only one sample, with ratings of 0 to 2 in the other four categories. The CC of *c* makes it clear that more Se data are needed on canned crab.

Table 4. Data quality evaluation for white rolls, whole milk, and canned crab

Description	Ref. no.	Number of samples (actual)	Data quality criteria ratings				Quality index	Se (µg/100 g)		Comments
			Analytical method	Sample handling	Sampling plan	Analytical quality control		Mean	SD	
White rolls										
Rolls, white, soft, enriched	7	2(4)	2	2	3	1	2.0[a]	25.50	14.06	4 composites of 3 samples each analysed
Rolls, white, soft, enriched	7	2(4)	2	2	3	0	1.8[a]	21.00	4.69	
White rolls	23	2(4)	1	2	2	0	1.4[a]	29	9	No Q.C. documentation
Rolls, brown and serve	3	2(3)	2	2	0	0	1.2[a]	61	24	Canadian
						$\Sigma =$ Mean Min.–Max.	$\underline{6.4}$ = =	34 21–61		Confidence code = *a*
Whole milk										
Whole fluid milk	7	2(4)	2	2	3	1	2.0	0		All values < QL
Whole fluid milk	7	2(4)	2	2	3	0	1.8	0		All values < QL
Whole or skim milk	23	2(9)	0	2	2	0	0	5.9	0.8	No method validation
Whole milk	23	3(48)	0	2	1	0	0	5.9	1.2[b]	No method validation
Whole milk	3	2(3)	3	2	0	0	1.4[a]	1.5	0.2	Canadian
Skim or whole milk	22	3(24)	0	2	2	0	0	6.9	2.1	No method validation
Whole homogenized milk	14	1(1)	2	2	2	0	1.4[a]	1.2		Duplicates

	No. of samples	Ref.					Index	Se value	
Whole fresh milk	15	1(1)	0	2	0	0	0.8	1.0	Duplicates; no method validation
Whole fresh milk	15	1(1)	0	2	0	0	1.1	1.7	
Regular milk (>3.5% fat)	4	1(?)	3	3	2		1.8^a	2.47	103 samples, all types duplicates; Canadian
Whole homogenized milk	2	1(1)	2	2	0		1.0^a	1.1	
Whole milk	10	1(?)	0	0	0	0	0	1.0	Reference is abstract
Milk	11	3(67)	0	0	1	0	0	4.0	No documentation
Whole milk	1	1(?)	0	0	1	0	0	4.8^b	Packaged in SD (no doc.)
Whole milk	1	1(?)	0	0	1	0	0	1.9^b	Packaged in OR (no doc.)
							$\Sigma =$	5.6	
							Mean =	1.6	Confidence code = *b*
							Min.–Max. =	1.1–2.5	
Canned crab									
Geisha King crab, canned	24	1(?)	1	0	2	0	0.8	51	No sample handling documentation duplicates
Crab, processed, packed in water	5	1(?)	1	2	2	2	1.2^a	21.68	
							$\Sigma = 0$		
							Mean	1.2	Confidence code = *c*
							=	22	

a. An index > 1.0 is required for inclusion of an individual Se value in the calculation of the mean.

b. Conversion from volume to weight based on item no. 01-077, "Whole Milk," in US Department of Agriculture, *Agriculture Handbook No. 8-1* (Science and Education Administration, USDA, Washington, D.C., 1975).

DISCUSSION

The main purpose of evaluating nutrient data is to eliminate poor-quality data, leaving only reliable information for the calculation of a mean value to be used in tables and data bases. A CC of *a* can mean that only three studies, two of them excellent, have been published or that a great deal of minimally acceptable data exists. In the case of Se data in general, a CC of *a* signifies the second situation. Most of the studies reporting the Se content of foods were assigned QIs between 1.0 and 2.0 out of a possible 3.0. A CC of *c* implies few data of minimally acceptable quality exist. A CC of *b* indicates data quality falling between *a* and *c*.

Combining and interpreting data from different studies presents some unique challenges to the nutrient data evaluator and statistician. Specifically, the biases of each study must be taken into account: biases based on different samples, analytical method, reagents, instrumentation, analysts' performance, and degree of accuracy and precision for each study. Usually these biases for a given study are not quantified or documented. Differences in the mean values for various studies cannot easily be evaluated when laboratories analyse samples obtained from different sources, and use different handling techniques, reagents, etc. The calculation of a mean nutrient value across studies can be performed in several ways. Weighting strategies were of particular interest. Weighting the mean towards the number of samples in the studies is one approach that was considered: data from studies that reported analyses of the largest numbers of samples would be weighted most heavily. However, this approach would attach greater significance to the number of samples category than to the others. Another approach might be to weight most heavily data from those studies with the smallest variance. However, this is not always possible because a standard deviation or standard error is sometimes not reported. A third approach can weight most heavily data from the study with the highest QI. This was also rejected owing to the narrow range of QIs and the subsequent lack of resolution in the scale. In view of the limitations discussed, weighting was deemed undesirable, and it was agreed to calculate a simple mean Se value at this time. As documentation and data quality improve, a weighting strategy could be considered for calculation of the mean nutrient value.

The derivation of the QI and consequent CCs also can be approached in several ways. A conservative scheme would be based on the assumption that the quality of a study is only as great as its weakest aspect, as was the system in the iron table [8]. From this viewpoint, the QI for each study would be equal to the lowest of the five ratings. However, applying this method of scoring to the existing Se studies would have resulted in very few acceptable data, since 0 is a frequent rating, especially in analytical quality control. Also, making the QI equivalent to the lowest rating would weight the QI toward the category with the greatest number of zero ratings. To avoid these consequences, a less conservative approach was taken which considered that: (a) sometimes quality-control measures are taken during the course of research, but not reported; and (b) the actual values found in Se papers with no mention of quality control often fall within or close to the range of values reported in Se papers that report appropriate quality-control measures. This holds true for the examples shown here. For the purpose of having enough acceptable data, standards have been adjusted. However, considering this compromise on the derivation of the QI, one safety feature was added to the calculation of the mean Se value: the exclusion of values with an index smaller than 1.0. This feature requires a study to meet a minimum level of overall quality for its data to be included. The minimum acceptable QI was set at 1.0 because that seemed to be a reasonable cut-off point, i.e. a higher cut-off point would eliminate the majority of studies.

Users of these data should be aware that the mean Se value for each food may not be representative of average levels found in the nation's food supply. Acceptable mean values were derived from available data from one or several studies. Individual criteria were not weighted, and even low ratings for sampling plan would not disqualify a study, depending on the other ratings. However, in each case the mean value represents the best present estimate of Se in a given food item.

IMPLICATIONS

One of the purposes of developing this system was to encourage investigators to consider all five categories of criteria when designing studies and reporting results. The system is a dynamic one and can be modified to respond to improvements in such areas as analytical methodology and availability of SRMs as well as to the reporting of new data. As additional research is done that incorporates the top levels of these criteria, we expect to upgrade our standards to allow more stringent evaluation of published data, and thereby increase users' confidence. For example, the quantitation limit of the method was rarely reported in the studies we evaluated. Without this information, it is difficult to assess the validity of low Se levels in foods. Rating studies on this aspect of analytical method only when a quantitation limit was reported was a compromise based on the level of existing data. In future evaluations, it is hoped that a rating of 0 will be assigned in analytical method to those studies that do not report the quantitation limit of the method as well as to studies which report results below the stated quantitation limit.

Although the criteria were developed using Se as the test case, they are applicable to data compilations for other food components. The evaluation system becomes nutrient-specific with the customizing of the criteria and the scheme for deriving the QI and mean nutrient value. Use of this evaluation system for any particular food component requires these steps: collection of relevant papers; delineation of nutrient-specific criteria at the various rating levels; assignment of ratings; and selection of strategies for deriving the QI and mean nutrient value. The quality of a given set of data influences the fitting of the criteria into the rating scale and the scheme selected for deriving the QI and mean nutrient value. This process is analogous to the familiar statistical problem of balancing type I and type II errors. If the rating scale is too rigorous, most available data will be eliminated. On the other hand, if it is too lenient, many less reliable studies will be included.

The delineation of levels of data quality permits data users to evaluate the suitability of a specific mean nutrient value for their data bases. Furthermore, access to criteria ratings that were assigned in the data evaluation procedure would allow the user of the data to assess the specific decisions made in the evaluation. The number of commercially available data bases is increasing, as are the number of users. Data users must take on the responsibility of selecting nutrient data of known quality.

ACKNOWLEDGEMENTS

The authors appreciate Dr Nancy Miller-Ihli's input on technical issues relative to analysis of foods, and gratefully acknowledge the statistical advice of Dr Estelle Russik and Dr Arnold Greenland.

DISCLAIMER

Names of products are included for the benefit of the reader and do not imply endorsement or preferential treatment by USDA.

REFERENCES

1. W. H. Allaway, J. Kubota, F. Losee, and M. Roth, "Selenium, Molybdenum, and Vanadium in Human Blood," *Arch. Environ. Health*, 16: 342–348 (1968).
2. M. A. Amer and G. J. Brisson, "Selenium in Human Food Stuffs Collected at the Ste-Foy (Quebec) Food Market," *J. Inst. Can. Sci. Technol. Aliment.*, 6: 184–187 (1973).
3. D. Arthur, "Selenium Content of Canadian Foods," *Can. Inst. Food Sci. Technol. J.*, 5: 165–169 (1972).
4. J. C. Bruhn and A. A. Franke, "Trace Metal and Protein Concentrations in California Market Milks," *J. Food Protect.*, 40: 170–173 (1977).
5. C. J. Cappon and J. C. Smith, "Chemical Form and Distribution of Mercury and Selenium in Edible Seafood," *J. Anal. Toxicol.*, 6: 10–21 (1982).
6. W. G. Cochran, *Sampling Techniques* (John Wiley & Sons, New York, 1977).
7. Committee on Dietary Allowances, Food and Nutrition Board, Commission on Life Sciences, National Research Council, *Recommended Dietary Allowances*, 9th ed. (National Academy Press, Washington, D.C., 1980).
8. J. Exler, *Iron Content of Food*, Home Economics Research Report, no. 45 (Consumer Nutrition Division, Human Nutrition Information Service, USDA, Washington, D.C., 1983).
9. S. N. Ganapathy, B. T. Joyner, D. R. Sawyer, and K. M. Hafner, "Selenium Content of Selected Foods," *Trace Elem. Metab. Man. Anim. Proc. Int. Symp. 3rd*, 1978: 322 (abstr).
10. D. M. Hadjimarkos and C. W. Bonhorst, "The Selenium Content of Eggs, Milk, and Water in Relation to Dental Caries in Children," *J. Pediatr.*, 59: 256–259 (1961).
11. L. H. Keith, W. Crummett, J. Deegan, Jr., R. A. Libby, J. K. Taylor, and G. Wentler, "Principles of Environmental Analysis," *Anal. Chem.*, 55: 2210–2218 (1983).
12. W. R. Mabey, J. S. Winterle, T. Podoll, et al., "Elements of a Quality Data Base for Environmental Fate Assessment, Final Report," Environmental Protection Agency Contract no. 68-03-2981, SRI Project PYU 2073, Work Assignment no. 6, 5 July 1984.
13. V. C. Morris and O. A. Levander, "Selenium Content of Foods," *J. Nutr.*, 100: 1383–1388 (1970).
14. A. L. Moxon and D. L. Palmquist, "Selenium Content of Foods Grown or Sold in Ohio," *Ohio Report*, 65: 13–14 (1980).
15. National Bureau of Standards Certificate of Analysis, Standard Reference Material 1577a, Bovine Liver (National Bureau of Standards, Washington, D.C., 1982).
16. National Bureau of Standards Certificate of Analysis, Standard Reference Material 1549, Non-Fat Milk Powder (National Bureau of Standards, Washington, D.C., 1984).
17. National Bureau of Standards Certificate of Analysis, Standard Reference Material 1571, Orchard Leaves (National Bureau of Standards, Washington, D.C., 1977).
18. National Bureau of Standards Certificate of Analysis, Standard Reference Material 1566, Oyster Tissue (National Bureau of Standards, Washington, D.C., 1979).
19. National Bureau of Standards Certificate of Analysis, Standard Reference Material 1568, Rice Flour (National Bureau of Standards, Washington, D.C., 1978).
20. National Bureau of Standards Certificate of Analysis, Standard Reference Material 1567, Wheat Flour (National Bureau of Standards, Washington, D.C., 1978).

21. O. E. Olson and I. S. Palmer, "Selenium in Foods Consumed by South Dakotans," *Proc. S.D. Acad. Sci.*, 57: 113–121 (1978).
22. O. E. Olson, I. S. Palmer and M. Howe, Sr., "Selenium in Foods Purchased or Produced in South Dakota," *J. Food Sci.*, 49: 446–452 (1984).
23. J. A. T. Pennington, B. E. Young, D. B. Wilson, R. D. Johnson, and J. E. Vanderveen, "Mineral Content of Foods and Total Diets: The Selected Minerals in Foods Survey, 1982–1984," *J. Amer. Diet. Assoc.* (in press).
24. H. A. Schroeder, D. V. Frost, and J. J. Balassa, "Essential Trace Metals in Man: Selenium," *J. Chron. Dis.*, 23: 227–243 (1970).
25. K. K. Stewart, "Nutrient Analyses of Food: A Review and a Strategy for the Future," in G. R. Beecher, ed., *Nutrition Research*, BARC Symposium no. 4 (Allanheld, Osmun & Co. Publishers, Totowa, N.J., 1981), pp. 209–220.
26. K. K. Stewart, "The State of Food Composition Data: An Overview with Some Suggestions," *Food Nutr. Bull.*, 5: 54–68 (1984).
27. S. Williams, ed., *Official Methods of Analysis of the Association of Official Analytical Chemists*, 14th ed. (Association of Official Analytical Chemists, Washington, D.C., 1984).
28. W. R. Wolf and M. Ihnat, "Evaluation of Available Certified Biological Reference Materials for Inorganic Nutrient Analysis, in W. R. Wolf, ed., *Biological Reference Materials: Availability, Uses, and Need for Validation of Nutrient Measurement* (John Wiley & Sons, Inc., New York, 1985), pp. 89–105.
29. US Department of Agriculture, *Agricultural Statistics* (US Government Printing Office, Washington, D.C., 1983).

19

Consideration of Food Composition Variability: What Is the Variance of the Estimate of One-day Intakes? Implications for Setting Priorities

G. H. BEATON

Department of Nutritional Sciences, Faculty of Medicine, University of Toronto, Toronto, Canada

INTRODUCTION

In another setting the author has been concerned with approaches to the nutritional assessment and interpretation of population data [6]. Two problems have been identified in this work: day-to-day variation in intake must be taken into account or there can be potentially serious errors in the estimation of the prevalence of either inadequate or excessive intakes; and the nutritional adequacy of intake (or the risk of excess associated with detrimental factors) must be approached on a probability basis. With replicated observations the first of these problems may be addressed by statistical adjustment of the distribution following ANOVA to estimate the partitioning of variance. The second can be addressed by generating probability statements based upon the distribution of requirements among individuals [1, 10, 6].

In the course of developing specific approaches applicable to large-scale surveys, consideration was directed to the question of food composition "errors." Current USDA tables provide estimates of the standard error associated with the average content figures for individual foods [9]. Standard deviations can be derived. These appear very large, with coefficients of variation (CV) ranging from about 10 to 50 per cent depending upon the food and nutrient. At first it appeared that this error, whether real (i.e. methodologic) or simply due to the range of compositions that a particular sample of a class of food might have (biological variation), was so large that any approach to assessment might be in jeopardy. To examine this specific issue, some examinations of the predicted "error" of one-day intakes were undertaken. The results of these examinations and considerations of implications are presented below.

This paper was prepared during the course of deliberations of a National Academy of Sciences Committee [6]. Much of the material now presented is to be found also in the National Academy of Sciences report together with an expansion of the discussion and statistical modelling of the error terms and their impact on the estimation of the prevalence of inadequate intakes. The interested reader is referred to that report for more complete information. Appreciation is expressed to the National Academy of Sciences for permission to reproduce tables now published in the report.

Table 1. Variability of food composition as empirically estimated from USDA composition tables[a]

Nutrient	Cut-off point	CV range assumed	
		Below cut-off	Above cut-off
Protein	2 g/100 g	5–50	5–15
Calcium	20 mg/100 g	5–50	5–15
Iron	1 mg/100 g	5–65	10–30
Magnesium	10 mg/100 g	5–50	10–30
Sodium	100 mg/100 g	5–65	5–15
Zinc	1 mg/100 g	5–65	10–30
Thiamine	0.05 mg/100 g	5–50	10–30
Riboflavin	0.05 mg/100 g	5–50	10–30
Niacin	0.5 mg/100 g	5–65	5–15
Vitamin C	7.5 mg/100 g	5–50	10–30
Vitamin B$_6$	0.1 mg/100 g	5–50	10–30
Folacin	20 μg/100 g	5–65	10–30
Vitamin A	300 IU/100 g	5–65	10–30

a. CVs estimated from reported SE of the mean value presented in table and reported number of determinations for the food [9].

MAGNITUDE OF THE REPORTED VARIABILITY OF COMPOSITION

Table 1 presents some estimates of the standard deviation of food composition generated from the standard errors and number of assays presented in the new USDA data bases [9]. This tabulation is impressionistic rather than systematic – that is, an examination of some sample foods rather than a thorough examination of all foods was carried out in order to generate the estimates. Empirically it was observed that for each of the nutrients there was a distinction between the CV at low concentrations and at higher concentrations. This was probably due to the impact of methodologic error when concentration was low. In table 1, estimates are presented above and below arbitrary cut-off points to illustrate this phenomenon.

It can be seen that the very high CVs are associated with the low concentrations of nutrients. Above the arbitrary cut-off points the nutrients fall into two classes as far as CVs are concerned. These probably represent the general magnitude of the biological variability of food composition.

IMPACT OF COMPOSITION VARIATION ON A ONE-DAY FOOD INTAKE

To assess the impact of this variation on the estimate of intake of an adult for one day, two sample diets, identified in tables 2 and 3 as HW1 and HW2, were taken from recorded food-intake studies. (The compositions of these diets can be found in the National Academy of Sciences report [6].) For each food and nutrient in these diets the CV of composition was either generated from the food composition table or, if the table presented no SE estimates, it was imputed, selecting a random value from the ranges suggested in table 1. Thus, for every food and nutrient in the two

Table 2. Food composition and variability estimates associated with food record HW1[a]

Food item	Weight eaten (g)	Composition per 100 g (percentages given in parentheses)												
		Protein (g)	Calcium (mg)	Iron (mg)	Magnesium (mg)	Sodium (mg)	Zinc (mg)	Vitamin C (mg)	Thiamine (mg)	Riboflavin (mg)	Niacin (mg)	Vitamin B_6 (mg)	Folate (μg)	Vitamin A (IU)
Orange juice	124.0	0.68 (0.6)	9.0 (15.3)	0.10 (64.1)	10.0 (8.9)	1.0 (158.3)	0.05 (43.4)	38.9 (13.5)	0.079 (21.9)	0.018 (0.0)	0.202 (10.6)	0.044 (0.0)	43.8 (18.0)	78.0 (26.1)*
Scrambled egg	64.0	9.32 (14.9)*	74.0 (13.1)*	1.46 (29.3)*	12.0 (21.4)*	242.0 (8.3)*	1.10 (28.1)*	0.2 (28.3)*	0.061 (18.3)*	0.243 (18.7)*	0.066 (52.0)*	0.091 (21.2)*	35.0 (28.0)*	486.0 (14.3)*
Sausage	26.0	13.80 (13.4)	0.0 (–)	1.13 (15.4)	12.0 (18.8)	805.0 (7.5)	1.87 (14.3)	0.0 (–)	0.357 (13.7)	0.147 (11.7)	3.367 (9.2)	0.050 (29.4)	0.0 (–)	0.0 (–)
Shredded wheat	28.4	11.00 (3.5)	38.0 (15.8)	4.22 (12.7)	132.0 (15.0)	10.0 (93.0)	3.30 (14.0)	0.0 (–)	0.260 (11.7)	0.280 (53.5)	5.25 (7.7)	0.253 (9.6)	50.0 (16.2)	0.0 (–)
Milk, whole	122.0	3.29 (5.3)	119.0 (6.9)	0.05 (44.3)	13.0 (37.4)	49.0 (16.3)	0.38 (16.2)	0.94 (50.2)	0.038 (27.6)	0.162 (14.0)	0.084 (15.7)	0.042 (49.4)	5.0 (28.3)	126.0 (34.5)
Coffee whitener	15.0	1.00 (44.2)	9.0 (62.9)	0.03 (223.6)	0.0 (–)	79.0 (60.9)	0.02 (25.3)	0.0 (–)	0.0 (–)	0.0 (–)	0.0 (–)	0.0 (–)	0.0 (–)	89.0 (82.9)
Tomato soup	248.0	2.46 (7.2)*	64.0 (14.6)*	0.73 (58.5)*	9.0 (11.8)*	376.0 (11.4)*	0.117 (27.5)*	27.3 (11.2)*	0.054 (12.6)*	0.100 (16.8)*	0.613 (12.9)*	0.066 (18.6)*	8.4 (37.1)*	342.0 (11.0)*
Frankfurters	90.0	11.28 (5.2)	11.0 (2.0)	1.15 (29.3)	10.0 (9.8)	1,120.0 (8.4)	1.84 (14.0)	26.0 (25.3)	0.199 (33.2)	0.120 (22.1)	2.634 (13.6)	0.130 (30.6)	4.0 (32.5)	0.0 (–)
Cheddar cheese	28.0	24.90 (7.2)	721.0 (8.6)	0.68 (25.1)	28.0 (19.6)	620.0 (16.7)	3.11 (21.4)	0.0 (–)	0.027 (55.0)	0.375 (24.9)	0.080 (67.6)	0.070 (46.2)	18.0 (55.8)	1,059.0 (25.2)
Peach	87.0	0.70 (20.0)	5.0 (47.7)	0.11 (83.3)	7.0 (27.3)	0.0 (–)	0.14 (59.1)	6.6 (37.8)	0.017 (35.3)	0.041 (19.5)	0.990 (4.4)	0.018 (23.6)	3.4 (28.2)	535.0 (10.2)
Milk shake	313.0	3.96 (8.0)	146.0 (12.6)	0.10 (19.9)	12.0 (13.4)	95.0 (7.3)	0.39 (31.5)	0.0 (–)	0.030 (33.9)*	0.195 (25.3)*	0.146 (22.0)*	0.042 (44.3)*	7.0 (56.1)*	114.0 (18.1)*
Potato soup	248.0	2.37 (10.8)*	67.0 (14.8)*	0.22 (38.4)*	7.0 (34.2)*	428.0 (5.4)*	0.272 (55.0)*	0.5 (16.1)*	0.033 (34.0)*	0.095 (21.0)*	0.259 (19.0)*	0.036 (14.7)*	3.7 (30.2)*	179.0 (36.6)*
Pork chops	126.0	28.82 (16.3)	15.0 (21.8)*	0.81 (16.7)	30.0 (23.7)	67.0 (6.3)	2.38 (14.6)	0.3 (28.3)	0.894 (32.6)	0.323 (20.4)	5.231 (11.9)	0.400 (19.0)	9.0 (38.3)*	6.0 (6.2)*
Fruit cocktail	122.0	0.42 (7.4)	5.0 (9.0)	0.25 (40.4)	7.0 (8.8)	4.0 (48.4)	0.09 (29.4)*	2.1 (33.6)	0.016 (19.8)	0.011 (–)	0.363 (21.9)	0.052 (29.9)*	0.0 (–)	250.0 (23.2)
Skim milk	245.0	3.41 (4.3)	123.0 (13.3)	0.04 (38.2)	11.0 (17.0)	52.0 (27.1)	0.40 (36.9)	0.98	0.036	0.140	0.088	0.040	5.0	204.0
Number of items with imputed CV (*)		3	4	3	3	3	5	3	4	4	4	5	5	6

a. CV values are shown below average composition.

196

Table 3. Food composition and variability estimates associated with food record HW2[a]

Food item	Weight eaten (g)	Composition per 100 g (percentages given in parentheses)												
		Protein (g)	Calcium (mg)	Iron (mg)	Magnesium (mg)	Sodium (mg)	Zinc (mg)	Vitamin C (mg)	Thiamine (mg)	Riboflavin (mg)	Niacin (mg)	Vitamin B6 (mg)	Folate (μg)	Vitamin A (IU)
Grape juice	126.0	0.56 (38.2)	9.0 (22.4)	0.24 (56.7)	10.0 (7.9)	3.0 (41.9)	0.05 (16.0)	0.10 (36.4)*	0.026 (51.0)	0.037 (46.8)	0.262 (13.9)	0.065 (10.2)	2.6 (21.1)	8.0 (37.1)*
Farina, cooked	117.0	1.4 (11.3)*	2.0 (45.3)*	0.50 (40.4)*	2.0 (15.4)*	0.0 (–)	0.07 (41.4)*	0.0 (–)	0.08 (27.5)*	0.05 (11.8)*	0.55 (5.9)*	0.010 (15.5)*	2.0 (12.0)*	0.0 (–)
Cream (half and half)	121.0	2.96 (5.8)	105.0 (9.2)	0.07 (25.7)	10.0 (13.5)	41.0 (7.1)	0.51 (49.2)	0.86 (12.2)*	0.035 (30.0)*	0.149 (28.1)*	0.078 (38.4)*	0.039 (45.5)*	2.0 (26.0)	434.0 (15.6)*
Fried egg	46.0	11.7 (5.7)*	56.0 (5.1)*	2.01 (23.7)*	12.0 (17.4)*	312.0 (11.8)*	1.38 (21.8)*	0.0 (–)	0.071 (20.9)*	0.275 (12.3)*	0.057 (45.7)*	0.109 (17.1)*	47.0 (27.1)*	622.0 (10.6)*
Lemon juice	2.5	0.38 (25.9)	7.0 (19.6)*	0.03 (13.3)*	6.0 (33.2)	1.0 (68.9)	0.05 (61.9)*	46.0 (20.4)*	0.030 (32.0)*	0.010 (23.8)*	0.100 (24.0)*	0.051 (11.6)*	12.9 (1.8)	20.0 (10.4)*
Yoghurt (low fat)	227.0	5.25 (10.9)	183 (11.7)	0.08 (20.0)	17 (11.8)	70 (38.5)*	0.89 (38.9)	0.80 (70.7)	0.044 (19.7)	0.214 (13.2)	0.114 (50.0)*	0.049 (53.6)	11.0 (23.6)	66.0 (27.7)*
Vegetarian vegetable soup	241.0	0.87 (13.2)*	9.0 (32.9)*	0.45 (58.4)*	3.0 (6.9)*	341.0 (8.7)*	1.191 (33.8)*	0.6 (21.8)*	0.022 (47.3)*	0.019 (5.5)*	0.380 (54.9)*	0.023 (46.4)*	4.4 (9.5)*	1,247.0 (18.8)*
Purple plums (canned)	129.0	0.36 (8.6)	9.0 (19.9)	0.84 (59.6)	5.0 (30.7)	19.0 (29.0)	0.07 (37.9)*	0.4 (39.7)	0.016 (16.5)	0.038 (27.9)	0.291 (12.6)	0.027 (5.8)*	2.5 (65.0)	259.0 (69.1)
Whole chocolate milk	250.0	3.17 (3.4)	112.0 (10.1)	0.24 (22.0)	13.0 (10.7)	60.0 (19.4)	0.41 (12.0)	0.91 (27.6)*	0.037 (46.2)*	0.162 (24.5)*	0.125 (49.2)*	0.040 (29.0)*	5.0 (62.6)*	121.0 (27.9)*
Turkey, roast	170.0	29.6 (2.2)	20.0 (15.5)	1.96 (22.0)	26.0 (1.6)	67.0 (8.6)	3.04 (4.1)	0.0 (–)	0.046 (16.0)	0.188 (13.0)	5.3 (4.4)	0.048 (23.8)*	8.0 (24.2)*	0.0 (–)
Cranberry sauce	69.0	0.20 (26.2)*	4.0 (37.2)*	0.22 (32.0)*	3.0 (30.5)*	29.0 (37.4)*	0.05 (63.7)*	2.0 (14.5)*	0.015 (47.1)	0.021 (0.0)	0.100 (46.4)	0.014 (0.0)	0.0 (–)	20.0 (17.6)*
Green pea soup	250.0	3.44 (12.1)*	11.0 (6.6)*	0.78 (57.6)*	16.0 (26.7)*	395.0 (5.9)*	0.68 (61.6)*	0.7 (5.2)*	0.043 (35.4)*	0.027 (36.0)*	0.496 (44.9)*	0.021 (36.3)*	0.7 (43.5)*	61.0 (56.7)*
Ice cream (vanilla)	133.0	3.61 (15.8)	132.0 (11.7)	0.09 (49.5)	14.0 (9.3)	87.0 (29.3)	1.06 (62.4)	0.53 (45.9)*	0.039 (20.4)	0.247 (18.8)	0.101 (14.7)	0.046 (9.7)*	2.0 (19.2)*	408.0 (20.0)
Number of items with imputed CV (*)		5	5	6	6	5	7	8	7	7	8	10	7	9

a. CV values are shown below average composition.

records, a mean and CV of composition was generated. The SD of composition was thus available as well.

A simulation operation was then conducted. For each food and nutrient item, a random composition value was obtained from the normal distribution described by the mean and CV developed as above. This was done independently for each food. The intake summed across all foods was then estimated. This exercise was repeated 1,000 times and then the mean and standard deviation of the 1,000 estimates of nutrient intake were calculated. This SD is a measure of the variability or error term associated with the one-day intake estimates that would be derived by using the average composition values presented in the food composition tables. The results of this exercise are shown in table 4. In comparison to table 1, there is a striking reduction in the relative variability of food composition in the mixed diet compared to that of the individual foods!

This can also be estimated by statistical formula. The variance of a sum is simply the sum of the variances. Applying this approach, the above results can be reproduced by statistical equation rather than by simulation analyses. Extending this approach, the relationship between the number of foods included in a diet and the variability of composition of individual foods can be presented in matrix form (table 5). It is to be noted that for the purposes of this table all foods are assumed to make an equal contribution to the total nutrient intake, which was not the case in the simulation analyses presented in table 4 or in other real diets. Table 5, then, must be seen as illustrating a principle.

The phenomenon portrayed is self-apparent: as the number of food items increases, the relative variability of composition decreases. That is, with a very diverse diet, variations or random errors in the composition of individual foods has a much smaller impact than in a diet with only a few foods. Diets HW1 and HW2 contained 15 and 13 foods respectively.

ADDITIONAL IMPACT OF A RANDOM ERROR IN INTAKE ESTIMATION

Given the above construct of the impact of food composition variation, in order to offer informational perspective it is possible to extend the model to include the impact of another source of variation – an error in the estimation of portion size when obtaining the record of intake. In this case the two variances (composition and intake estimate) would be multiplicative. Assuming that they are not correlated, the following equation would serve to predict the variance of the estimated nutrient content of a single food item. The equation is based on one presented in the FAO/WHO/UNU report on *Energy and Protein Requirements* [10].

$$V_{(food\ 1)} = I^2 * V_{(C)} + C^2 * V_{(I)} + V_{(C)} * V_{(I)}$$

where
- I^2 is the square of reported (mean) intake of units of food;
- C^2 is the square of reported (mean) concentration of nutrient per unit of food;
- $V_{(food\ 1)}$ is the variance of content of food 1 where content is $I*C$;

- $V_{(I)}$ is the variance of the intake measurement;
- $V_{(C)}$ is the variance of the composition measurement.

(The equation presented assumes no correlation between values of I and C; approximations are available for situations in which correlation exists.)

This equation derives the combined variance for a single food. To obtain the variance (and then SD and CV) for a one-day intake, the variances of the individual foods must be summed. The exercise was carried out first in a simulation exercise with 1,000 iterations and then by the statistical formula. The results were in very close agreement. For simplicity, table 6 presents the results of application of the above equation only for diets HW1 and HW2. The assumed error term in food-intake estimation had a CV of 10 per cent of the reported intake of the food item.

As was done above, this model can be generalized to illustrate the principles involved (table 7). To extend this to a one-day intake, it was assumed that all foods made an equal contribution to the one-day intake and that the diet contained 15 foods. Table 8 then presents the estimate of the variability that would be associated with the estimate of one-day intake.

If the data in table 8 are compared with the line in table 5 portraying the impact of food composition variation in a diet containing 15 foods, the impact of the additional source of error in the estimation of food intake can be seen. Thus, for example, if the composition variability is taken as 20 per cent and the intake error as 0 (table 5), the variation term in the one-day intake is 5.2 per cent. If an error term of 10 per cent CV in the intake estimation is now added (table 8), the one-day intake estimate has a variability of 5.8 per cent. The addition of the second error term has only relatively small impact!

SOME IMPLICATIONS FOR DATA ANALYSES

In assessment of the distribution of observed intakes, as in estimating the prevalence of inadequate or excessive intakes, the focus of attention should be on the estimation of the distribution of usual intakes. This is achieved by obtaining replicate one-day intakes. In the past these have been averaged for the individual to improve the estimate of his or her intake (eliminate the effect of day-to-day variation). A better approach involves application of an analysis of variance to estimate the distribution of usual intakes (the inter-individual variation component) [6]. Either way, a part of the random variation in food composition or in intake estimation will be factored out along with the removal of day-to-day variation. Thus, it can be demonstrated that the presence of variance affects the estimation of prevalence of inadequate intakes or of excessive intakes. However, the effect is surprisingly small! Improvement of the food composition data base by increasing replications will improve the confidence of the prevalence estimate, but the cost-effectiveness for this purpose needs to be examined very carefully.

Another application of computed nutrient intakes is in connection with epidemiologic studies – as in regression or correlation analyses of the relationship between observed intake and some biological outcome. It has been demonstrated by several authors that an error term in the estimation of usual intake can attenuate correlations and will bias regression slopes toward 0 if intake is used as the independent variable.

Table 4. Potential error associated with estimated one-day intakes

Nutrient	Diet HW1			Diet HW2		
	Mean	SD	CV (%)	Mean	SD	CV (%)
Protein	104.6	6.20	5.93	97.5	2.21	2.27
Calcium	1,540.2	80.77	5.24	1,135.2	61.31	5.40
Iron	8.03	1.19	14.85	10.4	1.66	16.00
Magnesium	250.1	15.70	6.28	222.4	13.04	5.86
Sodium	4,129.5	157.36	3.81	2,589.8	121.73	4.70
Zinc	11.6	0.909	7.85	13.3	1.64	12.33
Thiamine	2.10	0.375	17.92	0.715	0.076	10.59
Riboflavin	2.60	0.205	7.90	2.13	0.154	7.22
Niacin	15.9	0.908	5.72	13.5	0.879	6.53
Vitamin C	153.1	11.91	7.77	11.8	1.54	13.00
Vitamin B_6	1.45	0.136	9.37	1.43	0.210	14.62
Folacin	184.3	19.80	10.74	97.1	12.02	12.38
Vitamin A	3,798.4	281.24	7.40	5,142.0	603.61	11.74

Table 5. Impact of the number of food items in a record on the error term of computed nutrient intake for an individual[a]

Number of foods in record	CV of nutrient content of individual food serving									
	10	15	20	25	30	35	40	45	50	55
2	7.07	10.61	14.14	17.68	21.21	24.75	28.28	31.82	35.36	38.89
3	5.77	8.66	11.55	14.43	17.32	20.21	23.09	25.98	28.87	31.75
4	5.00	7.50	10.00	12.50	15.00	17.50	20.00	22.50	25.00	27.50
5	4.47	6.71	8.94	11.18	13.42	15.65	17.89	20.12	22.36	24.60
10	3.16	4.74	6.32	7.91	9.49	11.07	12.65	14.23	15.81	17.39
15	2.58	3.78	5.16	6.45	7.75	9.04	10.33	11.62	12.91	14.20
20	2.24	3.35	4.47	5.59	6.71	7.83	8.94	10.06	11.18	12.30
25	2.00	3.00	4.00	5.00	6.00	7.00	8.00	9.00	10.00	11.00
30	1.83	2.74	3.65	4.56	5.48	6.39	7.30	8.22	9.13	10.04

a. Table assumes that all foods make an equal contribution to nutrient intake.

A major concern in this regard is the day-to-day variation in intake [2, 3, 4, 5, 8], which can have a CV in the order of 25 to 35 per cent or even higher for some nutrients [2,3]. It is clear that food composition variation also contributes to the true total variability (it would be in *addition* to estimated intra-individual or within-person variation based on published food composition data). Clearly, improvement of the food composition data would improve correlation or regression analyses; it might not be very cost-effective. If we must continue to estimate food intakes on several days, and then average these to estimate "used intake," we again find that the random component of food composition variation will be reduced. The phenomenon is illustrated in table 9.

Table 6. Estimate of error term in one-day intakes associated with combined variability of food composition and error of the intake estimate (assumed measurement error CV = 10 per cent of reported intake, normally distributed)

Nutrient	Diet HW1[a]			Diet HW2[a]		
	Mean	SD	CV (%)	Mean	SD	CV (%)
Protein	109.6	7.56	7.23	97.5	5.81	5.96
Calcium	1,540.2	103.7	6.74	1,135.2	82.52	7.26
Iron	8.03	1.23	15.35	10.40	1.73	16.62
Magnesium	250.0	17.72	7.08	222.4	15.51	6.97
Sodium	4,129.5	239.3	5.80	2,589.8	180.3	6.95
Zinc	11.58	1.00	8.67	13.32	1.76	13.22
Thiamine	2.10	0.395	18.85	0.716	0.080	11.13
Riboflavin	2.60	0.226	8.71	2.13	0.175	8.21
Niacin	15.89	1.18	7.43	13.46	1.28	9.49
Vitamin C	153.1	14.78	9.65	11.85	1.61	13.56
Vitamin B$_6$	1.45	0.149	10.26	1.43	0.227	15.83
Folacin	184.3	21.12	11.46	97.07	12.72	13.10
Vitamin A	3,798.4	313.2	8.25	5,142.0	683.0	13.28

a. For composition variability, see table 5.

Table 7. Impact of random error in intake and food composition data on the error of calculated nutrient content of an individual serving of food[a]

CV1	CV1									
	0	5	10	15	20	25	30	35	40	45
0	0	5	10	15	20	25	30	35	40	45
5	5	7.1	11.2	15.8	20.6	25.5	30.5	35.4	40.4	45.3
10	10	11.2	14.2	18.1	22.4	27.0	31.8	36.6	41.4	46.3
15	15	15.8	18.1	21.3	25.2	29.4	33.8	38.4	43.1	47.9
20	20	20.6	22.4	25.2	28.6	32.4	36.6	40.9	45.4	50.1
25	25	25.5	27.0	29.4	32.4	35.9	39.8	43.9	48.2	52.7
30	30	30.5	31.8	33.8	36.6	39.8	43.4	47.3	51.4	55.7
35	35	35.4	36.6	38.4	40.9	43.9	47.3	51.0	55.0	59.1
40	40	40.4	41.4	43.1	45.4	48.2	51.4	55.0	58.8	62.8
45	45	45.3	46.3	47.9	50.1	52.7	55.7	59.1	62.8	66.8

a. All values expressed as CV = 100 × SD/Mean. It is not important which variable is 1 or 2.

VALIDATION OF FOOD INTAKE DATA: IMPLICATIONS OF FOOD COMPOSITION VARIATION

A particular consideration of the impact of food composition variation on regression and correlation analyses arises in connection with validation trials in which food intake estimated by recall or observation methods is compared with intake during the same period estimated by direct chemical measurement of duplicate meals. This is a common procedure. What may not always be recognized is that variation in food composition will inevitably yield a bias in regression slopes and attenuation of correlation coefficients even if the estimation of food intake is perfect. Further, the impact on regression depends on whether the error term lies in the dependent or independent variable.

Table 8. Estimate of variability (CV) of one-day intake derived from consideration of both food composition variation and random error in the estimation of intake (assuming 15 foods in the diet)[a]

	CV1							
CV1	10	15	20	25	30	35	40	45
10	3.7	4.7	5.8	7.0	8.2	9.5	10.7	12.0
15	4.7	5.5	6.5	7.4	8.7	9.9	11.1	12.4
20	5.8	6.5	7.4	8.4	9.5	10.6	11.7	12.9
25	7.0	7.6	8.4	9.3	10.3	11.3	12.4	13.6
30	8.2	8.7	9.5	10.3	11.2	12.2	13.3	14.4
35	9.5	9.9	10.6	11.3	12.2	13.2	14.2	15.3
40	10.7	11.1	11.7	12.4	13.3	14.2	15.2	16.2
45	12.0	12.4	12.9	13.6	14.4	15.3	16.2	17.2

a. Values calculated as value in table 7 divided by square root of 15, the assumed number of food items in the one-day diet.

This can be illustrated in a simulation analysis. Consider a model in which iron intake is computed from estimated food intake and is chemically determined from duplicate meals. Assume that for 97 individuals the intakes range from 5 to 25 mg per day and that the individuals are randomly distributed across this range. Consider also that the iron composition for one-day intakes has a CV of 15 per cent, a value consistent with the estimates presented in table 4. In the simulation analysis, 97 random values for iron intake lying in the 5 to 25 mg range were selected. For each of these mean intake values, a random value was selected from the population of possible real values described by a random distribution having mean as specified and CV = 15 per cent. The regression across the 97 individuals was then computed. This exercise was then repeated *de novo* 1,000 times. Finally the regression parameters for the 1,000 estimates were examined. The results are presented below:

A. With X = calculated intake (no error term in this model)
      Y = chemically determined intake (includes the "error" term)

      Regression parameters
        Intercept 0.0639 ± 0.5440[a]
        Slope 0.995 ± 0.0441
        Correlation coefficient 0.9229 ± 0.0146

B. With X = chemical composition (includes the "error" term)
      Y = calculated intake (no error term in this model)

      Regression parameters
        Intercept 2.1579 ± 0.5403
        Slope 0.8573 ± 0.0379
        Correlation coefficient 0.9229 ± 0.0146

a. Mean $\pm$ SD from 1,000 iterations of model.

That the bias in regression slope is seen only in one variation of the model is a recognized phenomenon. It is the error term in the independent variable that biases the regression.

Table 9. The impact of error in the independent and dependent variables on (A) regression slope and (B) correlation coefficient[a]

A. Impact on regression slope

Dependent variable (Y) variance ratio	Independent variable (X) ratio of intra/inter variances							
	0	0.4	0.8	1.2	1.6	2.0	2.4	2.8
0	1.0	0.714	0.556	0.455	0.385	0.333	0.294	0.263
0.4	1.0	0.714	0.556	0.455	0.385	0.333	0.294	0.263
0.8	1.0	0.714	0.556	0.455	0.385	0.333	0.294	0.263
1.2	1.0	0.714	0.556	0.455	0.385	0.333	0.294	0.263
1.6	1.0	0.714	0.556	0.455	0.385	0.333	0.294	0.263
2.0	1.0	0.714	0.556	0.455	0.385	0.333	0.294	0.263
2.4	1.0	0.714	0.556	0.455	0.385	0.333	0.294	0.263
2.8	1.0	0.714	0.556	0.455	0.385	0.333	0.294	0.263

B. Impact on correlation coefficient[b]

Dependent variable (Y) variance ratio	Independent variable (X) ratio of intra/inter variances							
	0	0.4	0.8	1.2	1.6	2.0	2.4	2.8
0	1.0	0.845	0.745	0.674	0.620	0.577	0.542	0.512
0.4	0.845	0.714	0.630	0.570	0.524	0.488	0.458	0.434
0.8	0.745	0.630	0.555	0.503	0.462	0.430	0.404	0.382
1.2	0.674	0.569	0.503	0.455	0.418	0.389	0.365	0.346
1.6	0.620	0.524	0.462	0.418	0.385	0.358	0.336	0.318
2.0	0.577	0.488	0.430	0.389	0.358	0.333	0.313	0.296
2.4	0.542	0.458	0.404	0.366	0.336	0.313	0.294	0.278
2.8	0.513	0.434	0.382	0.346	0.318	0.296	0.278	0.263

a. The reference slope and correlation are each set at 1.0. The tables portray the bias introduced as a multiple of true values.
b. All calculations presented assume that there is no correlation between errors.

A random error in the dependent variable has no specific effect on the slope. In the case of correlation analyses, the effect is the same no matter where the error lies. Here the correlations are high since the range of observed intakes is quite large in relation to the error term stipulated. More general models of these effects on regression slopes and on correlation coefficients are presented in table 9. The bases of these calculations will be found in Beaton et al. [2] and Snedecor and Cochrane [7].

SYSTEMATIC ERRORS IN FOOD COMPOSITION DATA

The effects described in table 9 and discussed in the foregoing text are to be distinguished from a bias in the estimates of average intake for the group of subjects. If statistically significant differences between estimated and determined intakes are found, it is suggestive of systematic bias in either the food composition table or in the estimation of food intake. (The latter may be dismissed if food composites have been based on reported intake rather than being true duplicate meals. Both approaches to validation of food intake have been used. Building food composites from reported

203

diets can be seen as a test of the validity of food composition tables. In this case the biological variability of individual samples of food should be considered in interpreting results.)

Any source of systematic error in food composition data will, of course, lead to a bias in the estimate of intake.

To consider the import of improving the food composition data base on data analyses involving regression or correlation analyses, it is necessary to consider concurrently the other error sources that may be present in the nutrient-intake estimates. It is necessary to recognize also that no matter how much the data base may be improved, there will always remain a biologic variability of composition among individual samples of foods. This will ultimately prove to be the limiting factor in improving the composition data base.

RELEVANCE TO PRIORITIES FOR FOOD COMPOSITION DATA

The above considerations have import in considering priorities in improving food composition data bases. Some of these are outlined below.

First, if there is any suspicion of true bias in the food composition data, the error will carry through all calculations. Thus, if there is suspicion that a methodologic error gives consistent under- or overestimation of composition, correction of the error should have high priority.

Second, it should be apparent from the foregoing that the major contribution to the error term in estimated one-day intakes will be associated with the foods that make the greatest contribution to total nutrient intake. That is, greater benefit will accrue from improvement of the composition estimate of the major nutrient contributors than from improvement of minor contributors. Therefore, obtaining more replicates of composition for major contributors will be more cost-effective than addressing minor contributors.

Similarly, and particularly in the connotation of the INFOODS programme, which will have international implications, the effect of improving the reliability of food composition will be greater in diets that include only a few foods than in diets that are marked by great diversity. It follows then that increasing the composition replicates will be more cost-effective in major foods of limited diets than in the case of diverse diets.

Obviously, if data are missing for certain foods, either imputed values must be used or intake from that food will be taken as 0. Either way there is a potential error. In the latter instance, the error will always be a bias toward underestimation of total intake; in the former, the error could be in either direction, across foods it might even be random. If the food is a major contributor of total nutrient intake, then the error term could be quite important. It follows that filling in missing data in the food composition table must have a priority. If the food is expected to be a minor contributor to total intake, and if reasonable imputations can be made, imputation of missing data may be quite reasonable. Conversely, if the food is thought to be a major contributor, it will be cost-effective to undertake analyses. Unless there is an a priori reason to believe that varietal differences are great, it may not be cost-effective to undertake composition determinations for each variety in use. The same consideration will apply to compositional differences attributable to soil composition and growing conditions (but note here that if the soil composition and growing conditions of a specific area affect all, or many, of the foods consumed in that area, a bias in the estimate of intake could be present). Perhaps the most cost-effective approach in this case would be research intended to determine whether major effects are likely to be present, and then a reconsideration of analytical priorities.

In all considerations, across nutrients, a scale of relative priorities must be based upon the perceived importance of examination of the nutrient in question as well as the relative cost of determinations.

CONCLUSIONS

If we are to address priorities for improvement of food composition data bases through increased numbers of analyses of food samples, the foregoing may be taken as an example of the merit of the following recommendation: Before major investment in food composition analyses is undertaken, either to increase the number of replicate analyses or to refine the specification of foods for which analyses are presented, it is recommended that the anticipated uses of food composition data be defined and "sensitivity analyses" be undertaken to examine the cost-effectiveness of such investment. From such sensitivity analyses a ranking of relative priorities, based on cost-effectiveness considerations, should be developed.

REFERENCES

1. G. H. Beaton, "Nutritional Assessment of Observed Dietary Intake: An Interpretation of Recent Requirement Reports," in H. H. Draper, ed., *Recent Advances in Nutrition Research* (in press).
2. G. H. Beaton et al., "Sources of Variance in 24-hour Dietary Recall Data: Implications for Nutrition Study Design and Interpretation," *A.J.C.N.*, 32: 2546–2559 (1979).
3. G. H. Beaton et al., "Sources of Variance in 24-hour Dietary Recall Data: Implications for Nutrition Study Design and Interpretation. Carbohydrate Sources, Vitamins and Minerals," *A.J.C.N.*, 37: 986–995 (1983).
4. D. R. Jacobs, Jr., J. T. Anderson, and H. Blackburn, "Diet and Serum Cholesterol. Do Zero Correlations Negate the Relationship?" *Am. J. Epidemiol.*, 110: 77–87 (1979).
5. K. Liu, J. Stamler, A. Dyer, J. McKeever, and P. McKeever, "Statistical Methods to Assess and Minimize the Role of Intra-individual Variability in Obscuring Relationships between Dietary Lipids and Serum Cholesterol," *J. Chronic Dis.*, 1: 399–418 (1978).
6. National Academy of Sciences, National Research Council, Subcommittee on Criteria for Dietary Evaluation, *Nutrition Adequacy: Assessment Using Four Consumption Surveys* (National Academy Press, Washington, D.C., 1985).
7. G. W. Snedecor and W. G. Cochrane, *Statistical Methods*, 7th ed. (Iowa State University Press, Ames, Iowa, 1980).
8. R. A. Stallones, "Comments on the Assessment of Nutritional Status in Epidemiologic Studies and Surveys of Populations," *A.J.C.N.*, 35: 1290–1291 (1979).
9. US Department of Agriculture, "Composition of Foods: Raw, Processed, Prepared," *Agriculture Handbook No. 8* (Science and Education Administration, USDA, Washington, D.C., 1976–). This reference is intended to cover a series of reports that is not yet complete.
10. WHO, *Energy and Protein Requirements*, report of a Joint FAO/WHO/UNU Expert Consultation, WHO Technical Report Series, no. 724 (WHO, Geneva, 1985).

20

Dietary Assessment Methods Used by the National Health and Nutrition Examination Surveys (NHANES)

CATHERINE E. WOTEKI

National Center for Health Statistics, US Department of Health and Human Services, Hyattsville, Maryland, USA

INTRODUCTION

The National Center for Health Statistics has conducted health examination surveys for over 20 years. The National Health Survey Act of 1956 authorized the secretary of what is now the Department of Health and Human Services, acting through the National Center for Health Statistics, to collect statistics on a wide range of health issues. Among other topics, the centre is authorized to collect statistics on "determinants of health" and "the extent and nature of illness and disability of the population of the United States (or of any groupings of the people included in the population) ..."

As part of its response to this mandate, the centre fielded the first National Health Examination Survey in 1959. The target population for this survey was adults of 18 to 74 years of age. Two additional surveys were conducted during the 1960s, extending the age groups examined to include children of 6 to 11 years and adolescents of 12 to 17. In 1971, the range of topics included in the survey was extended to include nutritional status. Nutritional status was assessed through a fivefold approach including a medical history, a physician's examination, biochemical tests, body measurements, and a dietary interview. The first National Health and Nutrition Examination Survey (NHANES I), conducted from 1971 to 1974, examined a representative sample of persons between the ages of 1 and 74. An additional sample of adults aged 25 to 74 years, called the NHANES I Augmentation, was examined in 1974–1975. The second National Health and Nutrition Examination Survey (NHANES II) was conducted from 1976 to 1980. The age range extended to include infants of six months to one year. In December 1984, the centre completed data collection for the Hispanic Health and Nutrition Examination Survey (NHANES) of persons of Mexican–American, Puerto Rican, and Cuban ancestry residing in the continental United States. We are beginning to plan the next survey, NHANES III, scheduled to begin in 1988.

Over the years, data generated by the health examination surveys have served a variety of uses. The surveys have provided estimates of the prevalence of characteristics or conditions in the American population. Normative or descriptive data have been

206

published on topics such as weight and stature. Both types of estimates permit the monitoring or measurement of changes in health and nutritional status over time through successive surveys. Problems of public health importance have been identified. The survey data have also been used to study the interrelationships of health and nutrition variables in the general population.

My purpose is to describe the National Health and Nutrition Examination Surveys, paying particular attention to the dietary intake data. Using NHANES II as a reference point, I will discuss design considerations, the major components of the survey related to nutrition, uses of the dietary data, and plans for future surveys.

DESIGN OF NHANES II

In approaching the design of NHANES I and II, many factors had to be considered. I will discuss here the design specifications that were considered in planning the most recent national survey, NHANES II.

NHANES II was planned to be a profitability sample of the civilian, non-institution-alized population of the United States for persons of 6 months to 74 years. Three subgroups in the population were of special interest for nutritional assessment because it was thought they were at higher risk of malnutrition: pre-school children (6 months to 5 years), the aged (60 to 74 years), and the poor (persons below the poverty level as defined by the US Bureau of the Census). These groups were oversampled to improve the reliability of the statistics generated. Although women of child-bearing age were also considered to be at risk of malnutrition, no oversampling was necessary. The total sample size desired was 21,000 examined persons, and the number of sample persons selected in each primary sampling unit (PSU) was to be between 300 and 600.

The data collection mechanism used in NHANES I was used again in NHANES II with appropriate modifications. An initial interview was conducted in the household in which socio-demographic information and medical histories were collected. Sample persons were scheduled to visit mobile examination centres in which the physical examination, dietary interview, anthropometry, and other procedures and tests were conducted. At any time during the survey period, two centres were operating in different locations while a third was being serviced or relocated. The mobile examination centres provide a controlled, standardized environment for the examinations and tests. The examinations and tests were conducted by a small, well-trained staff which moved from site to site with the mobile examination centres.

Because of the small number of mobile examination centres, the logistical constraints involved in moving and setting up the centres, the large number of sample persons, and the length of the examination, the total period for data collection was planned to be three to four years. The average length of an individual examination was two to three hours, but it varied depending on the age of the examinee. The examination for pre-school children lasted no more than two hours, while the time for an adult did not exceed three hours.

The survey was designed to produce statistics for the four broad geographic regions of the United States and for the total population by age, sex, race, and income classifications. In the end, a total of 20,322 individuals were interviewed and examined in NHANES II in 64 primary sampling units. Because not all individuals underwent all aspects of the interview and examination, appropriate non-response adjustments were made. These non-response adjustments bring the sum of the final weights into close alignment with the age, sex, and race estimates of the Census Bureau at the mid-point of the survey.

MAJOR NUTRITION-RELATED COMPONENTS OF NHANES II

The five major components of NHANES II were a household questionnaire, a medical history questionnaire, a dietary questionnaire, examination by a physician, and special procedures and tests. The household questionnaire consisted of questions about family relationship; age, sex, and race of family members; housing information; occupation, income, and education level of each family member; and participation in the food stamp programme and school breakfast and lunch programmes. Separate medical history questionnaires were used depending on the age of the sample person, one questionnaire for children of 6 months to 11 years and another for persons of 12 to 74 years. Both the household questionnaire and the medical history questionnaire were administered in the respondent's home.

When individuals arrived at the mobile examination centre, they were scheduled through the dietary interview, physician's examination, and special procedures and tests. The procedures and tests included body measurements for all, allergy tests for persons of 6 to 74, X-rays of persons of 25 through 74 except pregnant women (cervical spine, lumbar spine except women under 50 years, and chest), and urine and blood tests. From blood samples taken in the centre, a number of nutrition-related assays were done. These included serum albumin, serum vitamins A and C, serum lipids (cholesterol, triglycerides, and high-density lipoproteins), protoporphyrin, serum iron, total iron-binding capacity, serum zinc, and serum copper. Red cell folates, serum folates, serum ferritin, and serum vitamin B_{12} were determined on blood samples with abnormal complete blood count, haemoglobin, haematocrit, or MCV, and on a subsample of normals.

The dietary questionnaires consisted of a 24-hour recall, a food-frequency questionnaire, a dietary-supplement questionnaire, and specific questions on medication, vitamin, and mineral supplement usage. All interviews were conducted by trained interviewers who had at least a bachelor's degree in home economics.

In the 24-hour recall, respondents were asked to report all foods and beverages consumed on the previous day. Respondents estimated the size of the portions consumed by referring to food models. In addition to foods and portion sizes, interviewers asked about what time of day the food was eaten and its source. The time of day was coded as one of five ingestion periods: morning, noon, between meals, evening, or total day. The source of the food was coded as home, school, restaurant, or other.

Each food item was coded by the interviewer within 72 hours of the interview. The food code book developed for the survey contained five-digit food codes for approximately 2,500 food items. Each food item was identified by name (including brand names if appropriate), by whether it was raw, dry, or frozen, by how it was prepared, and, for mixed dishes without food codes, by major ingredients. A food composition data base updated from NHANES I was used to calculate the energy, vitamin, and mineral content of the reported foods. Modifications to the NHANES I data base included new data from USDA's revised *Handbook No. 8*, and food composition data from food companies on new products and brand-name products of unique formulation.

The food-frequency questionnaire elicited information about the consumption of 18 food groups over the previous three months. Frequency was given one of four possible codes: a whole number, never, less than once a week, or unknown. The interval at which the food was usually eaten was also given one of four possible codes: never, daily, weekly, or less than weekly. One question was asked about use of vitamin and mineral supplements, and one about how often the salt shaker was used at the table. Responses to this last question could be assigned to one of three codes: rarely or never, occasionally or seldom, frequently or always.

The dietary-supplement questionnaire contained questions about whether the respondent was on a special diet, what type, and for how long. One question asked about the possible use of nine medications in the previous week. These were commonly prescribed medications that might interfere with test results or affect interpretation of results. Another question related to problems preventing the respondent from obtaining needed groceries. The final question asked about trouble swallowing, pain, nausea and vomiting following eating, and loss of appetite.

The medication, vitamin, and mineral usage questionnaire requested specific information about brand name, manufacturer's name, and reason for using vitamin or mineral supplements and medications.

The quality of the dietary component was controlled at several levels. Before the survey began, the dietary interviewers were trained in interview techniques and in how to code the 24-hour recall. A manual was issued to each interviewer which described the procedures to be followed. Periodically, the forms were reviewed and evaluated, and instructions were issued to the interviewers to promote consistency. Interviewers exchanged coded 24-hour recall forms to check each other's work, and forms were also reviewed by the field staff before being forwarded to headquarters. At every location, each interviewer tape-recorded two interviews with randomly selected subjects. The recordings were evaluated at headquarters for adherence to procedures. Comparisons were made at headquarters of the mean values and frequency distributions by stand location and by interviewer to detect unusual results by location and systematic errors by interviewers. Foods for which no appropriate food codes existed were forwarded to headquarters for assignment of new code numbers.

USES OF DIETARY DATA

NHANES dietary data have been put to four types of uses: relating diet and demographic characteristics, relating diet and health characteristics, determining interactions of diet and nutritional status indicators, and tracking trends in diet and nutrient intakes over time.

In relating diet to the demographic characteristics of the population, the major question to be asked is: What are the food consumption patterns and nutrient intakes of subpopulations of the United States by such characteristics as age, race, sex, income, occupation, and education? The NHANES dietary data can answer questions such as: How do nutrient intakes and food consumption patterns of persons differ by level of education? What are the regional differences in consumption of certain food groups?

NHANES data have been used to relate the food consumption patterns and nutrient intakes of United States subpopulations to indicators of health status. Specific questions that have been addressed include: How do nutrient intakes compare with the Recommended Dietary Allowances and other dietary guidelines? What dietary patterns are associated with higher levels of tooth decay? What dietary and health variables are associated with iron-deficiency anaemia?

Examining interactions between nutrition-related variables, NHANES data can compare dietary intake, biochemical status, anthropometry, and presence or absence of health conditions. Questions that can be addressed by the data include: What are the relationships between dietary intake and biochemical status for persons who smoke, use vitamin/mineral supplements, or use oral contraceptives? Are those who take vitamins and other dietary supplements the ones who need them? Are subpopulations with high serum cholesterol and other evidence of cardiovascular disease consuming foods high in cholesterol and saturated fats?

Changes over time in food and nutrient intakes can be tracked and correlations made with health variables. Examples of questions that can be posed to the data include: What changes in obesity and diet have taken place in the last ten years? Are serum cholesterol values declining among men and women?

PLANS FOR FUTURE NHANES

The next National Health and Nutrition Examination Survey (NHANES III) is scheduled to begin in 1988. We have already begun planning for the survey. Among the topics being considered are the content, sample design, data processing, co-ordination with other surveys, addition of a longitudinal component, and the possibility of continuous monitoring of special groups.

The needs of government agencies, including the Food and Drug Administration, the Environmental Protection Agency, and the National Institutes of Health, and researchers in industry and academia will be considered. As the content of the survey is being developed, consideration will also be given to which topics should be considered core components. The core components would be administered to all sample persons while the non-core components would be administered to a subsample. In addition, the core components would be repeated in future surveys.

Suggestions for content will be solicited from federal agencies, the legislative branch, the public health and nutrition communities, researchers, foundations, and associations. A variety of mechanisms are being considered to gather recommendations from these groups including letters, meetings, and advertisement in journals.

Concurrent with decisions about survey content, preliminary decisions must be made about sample design, data processing, co-ordination with other surveys, addition of a longitudinal component or continuous monitoring of high-risk groups. Some of the questions to be answered include: whether it is feasible to include primary sampling units from the Health Interview Survey or the Nationwide Food Consumption Survey as primary sampling units in NHANES III; where automation can improve turn-around time, cut costs, and decrease errors; and whether NHANES III can use the same food composition data base that was used in the Nationwide Food Consumption Survey. It is conceivable that the dietary interview could be automated in NHANES III. Coding and edit checks would be accomplished during data entry while the interview is conducted. Changes could also be made to the current format, increasing the number of 24-hour recalls per person, for example.

CONCLUSION

The last 15 years have been a period of unparalleled interest in the relationship of diet to health. The dietary component of NHANES together with the clinical and biochemical assessment form a unique data set on a nationally representative sample of people. NHANES data have been used to monitor changes in health, nutritional status, and dietary intake over time. Interrelationships among dietary and health variables in the general population have been studied. NHANES III will continue to build on this information foundation.

A problem in planning the dietary component for NHANES III is that conflicting demands are being made. Regulatory agencies and researchers want more detail about the food people eat, how it is packaged and prepared, and what nutrients, additives, and toxic substances it contains. Demands for more rapid publication of data would lead us to simpler interviews with less detail about the foods consumed.

While this dilemma probably cannot be resolved immediately, we would like to hear discussion of the pros and cons of shortened, simplified interviews and data bases for use with NHANES. We would also like to hear recommendations on how to make our national surveys more compatible while extending their usefulness to policy-makers and researchers.

21
Systems Considerations in the Design of INFOODS

JOHN C. KLENSIN

Laboratory of Architecture and Planning and INFOODS Secretariat, Massachusetts Institute of Technology, Cambridge, Massachusetts, USA

INTRODUCTION

The International Network of Food Data Systems (INFOODS) was organized in 1982 as a global collaborative of people and organizations interested in working towards improving the amount, quality, and availability of food composition data. Currently it is focusing on the development of standards and guidelines for (a) the terminologies and nomenclatures used in describing foods and food components, (b) the gathering of food composition data, including sampling, assay, and reporting procedures, and (c) the storing and interchange of food composition data.

INFOODS is co-ordinated by a small secretariat, based at the Massachusetts Institute of Technology, which has responsibility for initiation, co-ordination, and administration of international task forces to work on specific problems. Additionally, this secretariat serves as a resource for, and clearing-house for information about, food composition activities around the world. INFOODS works with, and is organizing where necessary, regional groups throughout the world; these provide information and assistance for food composition work in their geographic areas. INFOODS presently is funded primarily by the United States government and administratively supported by the United Nations University.

It is generally assumed that the major product of INFOODS will be one or two integrated computer systems for nutrient and nutritional data. In terms of both technical problems and the requirements of different groups of users, that goal presents serious challenges. It is useful to review those challenges and the reasons why a different strategy may be in order.

Technical problems and user requirements may be seen as challenges because they involve questions for which we don't know the answers, as well as several for which, at this point, we probably do. The validity of our belief in the answers we have depends on whether certain analogies hold between systems for management, recording, and analysis of nutrition data and those for other types of data – especially statistical and social measurement data – and the scientific application of them. In addition, a recurring theme in systems design is that large systems usually involve complex choices to which there are few "correct" answers. Instead, there are many trade-offs in which the needs and preferences of one group are optimized at the expense of others. Making these choices explicitly and with an understanding of their implications, and remembering, far into the future, the reasons for the options chosen, tends to promote better systems that are both more internally consistent and consistent with their avowed goals. Inevitably, making explicit choices that are remembered, some of the decisions will turn out, as time passes, to have been wrong. As a result, one of the major challenges – almost a meta-challenge – is designing for damage containment to ensure that a few wrong decisions do not result in the total uselessness of the system or the need to rebuild it from scratch. An understanding of how the wrong decisions were arrived at contributes to containment of the damage.

One of the themes that is not important is the question of "personal" v. "large" computers as ends in themselves. There can be specific reasons for choosing smallish machines – cost, space, even the psychological advantage of being able to pull the thing's plug if it behaves offensively; and there are also some reasons for choosing large ones (or complexes of small ones) – economies of scale, the ability to retain large data bases of foods or consumption histories, and convenient sharing of information among scientists. But in discussing the reasons for choosing a machine we should not get involved in debate about the relative merits of small and large computers. It is especially important to avoid that debate because the use of some mixed strategies, in which different equipment is used for different operations, may be the best overall strategy given the present state of the art.

Before discussing the issues, challenges, and problems involved in trying to construct integrated systems, we should look at the question of why such systems should be considered. Small non-integrated systems have several advantages. They are typically cheaper to build and easier to maintain, and do not require large team efforts over a long period of time. Perhaps as important is one of the major discoveries of the microcomputer revolution – that considerable "friendliness" is a characteristic of machines that are not very capable. When capability is limited, it becomes possible to list all the commands, to list all the options, and to provide clear error messages that identify all choices. In other words, a message such as "No, you cannot type that answer, you must use one of the following three" is a reasonable and possible option. It is neither reasonable nor possible if there are tens of options to a particular command. Nor is it feasible to respond to an inquiry about what a command is called by listing all commands when there are several hundred from which to choose. The limited environments of small and unintegrated systems also tend to make them comparatively easy to document.

In this paper, large-scale systems are assumed to be groups of programs that provide a more or less common face to users, that permit free movement of data and intermediate results between different commands or other program components and analyses, and that let the user determine the order and content of both analyses and display formats. Such assumptions make the large-scale system a different type of object, rather than just a larger one, from most traditional program packages or packaged programs.

If one can figure out what is to be done with the data and what analytic and accessing capabilities are needed, it is often easily possible to design a collection of several medium-sized programs or small-scale systems for quite different purposes and users and having different interfaces, to operate from a single data base. In terms of the complexities of getting the data-base design right, that type of arrangement raises the same issues as the large-scale system, but is much easier from a software design standpoint. Also the individual programs may be much easier to get onto a small machine than a complete large-scale system would be. So that is one of the alternatives to be considered.

A potential advantage of large systems is that they should be able to provide a user with more flexibility. At their best, they contain a wider resource base – more tools that can be applied – for most situations. If designed well, they should have a longer life expectancy than smaller systems because they can be extended further and can be used in more innovative and creative ways, including ways not anticipated by their designers. Larger systems can support a wider variety of models and analyses, and consequently permit comparisons among techniques. Such additional analytic capabilities are usually supplemented by facilities for handling large or complex data sets that are beyond the capabilities of a small system.

Most of the issues raised in this paper apply to smaller systems as well as larger ones, but become much more important as systems become larger. The would-be developers of a large system must consider these issues in the early design stages to avoid severe problems later on. The major challenges are easily stated: planning and designing what the system is to do and how to implement it, and then testing those ideas. Also essential, although seemingly obvious, is that resources adequate to the task be available not only at the beginning but also over a long enough span to do the entire job. The best long-term strategies, which tend to focus on the building of tools and prototypes and the conduct of experiments before a final commitment is made to a strategy, tend to be poor short-term ones from the standpoint of sponsors or agencies looking for results. The fear of ending up with only a prototype when the resources run out has prevented many prototypes from being built, and as a result many problems have occurred in production that would have been easily identified and eliminated in prototype experiments. The resource issue will not be addressed here, except to note that tasks usually take much longer and cost much more than expected.

It is worth noting that a very large fraction of the time and cost overruns in computer-system building and programming can be attributed to a lack of clarity about what the goals are, what users are to be served, and what facilities are to be incorporated. Clear thinking, careful analysis of requirements and constraints imposed by the users (as distinct from ones imposed by real or imagined technical considerations), and careful design consistent with that thinking and analysis are usually richly rewarded, and the failure to perform such thinking and analysis is equally richly punished.

STAFF TURNOVER AND SYSTEM GROWTH

The planning of large systems requires consideration of a future in which most of the members of the development group will change by the time the system is in active and productive use. By the time the system is ready for demonstration, many of the development staff will have departed, although the designers may well still be around. This implies that careful attention must be paid to how additions and modifications to the system will be made and how the system will be extended in the future either by the users or the design group. With the typical system design, there are benefits from building special tools to aid in system construction, integration, and testing. It is often useful to expend some effort to define and delimit the framework of the proposed system – its boundaries, fundamental structure, and relationship to the outside world. How much time and effort can and should be spent in these areas becomes another critical choice. This choice is complicated by the knowledge that what is appropriate for a central staff to do in developing a system may not be appropriate for a staff later on (especially one that is administratively or geographically dispersed) and may not be appropriate when users try to create their own extensions.

DOCUMENTATION

The question of how to document a large system is a key one that should be addressed early and as part of the planning of the code and user interfaces. One approach is to provide comprehensive documentation; but comprehensive documentation may run into thousands of pages as the system grows [17]. Such volume will almost certainly lead to complaints about size and bulk, comments about needing wagons rather than binders, and requests that everything be distilled onto cards than can be put into pockets and purses. Standards about information to be included in documentation – algorithms [7, 12, 16], error messages, and the like, as well as sampling information and methods of analyses – make such volume inevitable; a mere four pages of description on each of 500 commands leads to 2,000 pages of documentation. On the other hand, documenting a large system as if it were a small one, adopting pocket cards or brief on-line files as the only form of documentation, or in some other way trying to keep the total under 100 pages will cause user frustration or worse. These are questions that do not have clear answers, but making choices early and clearly and remembering the reasons for the decisions that are made can help. At the same time, these decisions, like others discussed throughout this paper, should be made in a way that minimizes the damage if the world appears differently at some time in the future.

THE CHOICE OF ENVIRONMENTAL AND BASIC TOOLS

Almost any applications system that one might build today will exist in some environment over which the system developers do not have complete control. The days of writing codes in absolute binary and keying them in from the front panels of machines have departed, some recent excesses in the microcomputer community notwithstanding. Potentially, this means that choices must be made about what environments will be established to develop and operate the system – choices about hardware, operating systems, and languages. In many cases, the possible choices are so constrained by circumstances as to be trivial or non-existent. Worse, the constraints often arise from circumstances that have nothing to do with the requirements of or the intentions of the new integrated system, and will often lead to choices that are pathological for it. For the reader who has the luxury of making choices the two sections that follow are provided; for the reader who does not, these sections may be helpful in anticipating problems where choices are more constrained.

CHOICES OF OPERATING SYSTEMS

In an ideal world, the operating system chosen for any applications system is one that is smart, flexible, and state-of-the-art, and that operates on powerful, inexpensive, widely available hardware. In addition, the operating system must be utterly stable, so applications development does not involve aiming at a moving target. These attributes almost never exist in combination. Advanced and state-of-the-art systems are typically kept that way by continual revision or frequent releases. Each revision will "improve" the environment in ways that more or less significantly undermine existing work. Applications system developers can gain control over such changes by developing and maintaining their own operating systems, but the price for doing so is usually too high. Systems should be selected to reach a reasonable balance between sophistication and modernness on the one hand and stability on the other. Once the selections are made, software design criteria should include the ability to keep the stable interface that end-users will insist upon; for once someone gets used to a system that is even moderately satisfactory it is likely to be strongly preferred to any other, even those that are objectively better. It will be expected that stability will be preserved even when the supporting operating system is changed.

CHOICE OF PROGRAMMING LANGUAGE

The choice of a programming language (or set of languages) usually follows that of an operating system. While there have been cases in which an operating system was chosen because it supported a particular language, such cases are rare. All things being equal, systems that can be built entirely in a single language are much easier to cope with than those that require two or three. If nothing else, use of a single language makes the management task easier, since few programmers are equally comfortable and efficient in several languages at the same time; just as with natural languages, it is difficult to "think" in more than one language concurrently, regardless of what one can manage at separate times or in different places.

Unfortunately, from the language standpoint, operating systems requirements involve procedures for data entry and recording, for locating and aggregating data, and for doing statistical computations on surveys and to construct food tables. Historically, almost all good software for screen management and data entry support has been created in or for COBOL or PL/I. By contrast, almost all of the research and development work in numerical algorithms has been done in FORTRAN and ALGOL-60. While FORTRAN and ALGOL are quite suitable for computational codes, they are not suitable for systems work unless machine dependencies, assembly language subroutines, and other forms of idiosyncratic and incomprehensible code are introduced. COBOL is terrible for systems work, and not much better for numerical computation. The languages that are very good for systems work tend to be too poor or untested for serious computational work. While there are two possible exceptions, both are very large and complex as languages go, and there are allegations that they are very clumsy and hard to learn; further, they tend not to appear very often (at least in complete form(in microcomputer implementations.[1] The alternative, writing in assembly languages rather than relatively high-level ones, usually leads to trouble, and should not be seriously considered in building a large system with a long life expectancy. There is a third alternative in languages like BCPL, BLISS, and C that are really medium-level, nearly machine-independent assembly languages. For serious applications work, however, as distinct from systems work, they can be nearly as much trouble as assembly languages and for much the same reasons.

USER INTERFACE

The user interface for a system includes not only how users will communicate with the system, but also how the system will communicate with users in normal and error situations and how output will be formated and presented. There are many opinions about each of these issues, and none is completely correct for all audiences. Every user interface decision is problematic. This paper cannot hope to provide a complete discussion of the issues; the examples that follow are intended to convey the flavour and difficulty of the challenge.

If a single user-level language is chosen and imbedded in the system, the choice must be correct for all present and future audiences. Interfaces that adapt automatically to the characteristics of individual users and their growing knowledge of the system are a major research area today. Where such adaptation is possible, it complicates documentation both for the users themselves and for those who are expected to understand the processing and analysis activities of others. If one can design language for the system around the particular needs of the users to be served, be they the builders of food tables, epidemiologists, or hospital dieticians, much convenience may be gained for the users, and their learning and use of the system may be expedited. At the same time, such language design may require a great deal of learning on the part of users with other backgrounds or interests.

If the needs of several different types of users are to be supported, the system will end up with multiple languages and interfaces and all of the inconsistency and unpredictability to which that leads. Since large, diverse systems attract people with diverse needs and backgrounds, there are no easy solutions – but this may be another argument for not building such systems at all.

One popular solution, at least hypothetically, is to try to use a natural language, such as English, as the means of instructing the system. With advances in the technology in the last few years, this is probably a feasible option, although it entails considerable difficulty in design and implementation. More significant difficulties lie with the degree to which natural languages lack compactness of form and absolute clarity, which is why mathematical and other symbolic notation is used in statistical work. One of the greatest difficulties in using natural languages is getting people to understand them unambiguously.

---

1. The two languages are PL/I and the US Department of Defense's new high-level language, Ada (Ada is a registered trademark of the US Department of Defense). The pros and cons of each are highly controversial in the computer science community, and are beyond the scope of this paper. The claim that they might be useful for both machine-independent systems work and for computation is based on the explicit design criteria for both languages, which include that type of mixed application use. PL/I is much more widely implemented than most people realize, but, while data-base and screen (forms) management programs have been developed for it, few statistical or numerical algorithms have been (algorithms from ALGOL-60 translate rather easily). Complete implementations of Ada have not yet, as of this writing, been tested in serious production computational use, regardless of the impressions that the popular computer press may be creating.

All of the preceding comments on the user interface apply to picture languages, pointing languages, and even shouting-at-the-computer languages just as much as they apply to typed commands. One can either optimize for a particular group of users and leave everyone else somewhat inconvenienced and unhappy, or one can try to find compromises somewhere.

The actual mode of communication between user and system is almost a separate issue from that of the language used in the communication process, although some choices in this area can constrain, or remove constraints from, language choices. User-oriented menus, help systems, command completion systems, and question-answering are good choices for raw beginners; it has been argued elsewhere [15] that they have a tendency to grow pathological for regular and experienced users. Many of the most interesting and sophisticated of the less traditional approaches to human–computer interfaces rely on specialized hardware which may prevent their application in many specific implementations.

One useful alternative to a single, firm, choice of interface is to design "agent" facilities so that the system can easily support programs – both system- and user-provided – that run other programs on behalf of the user. Agents may be useful for supporting alternative command formats and presentations, default arrangements, and a choice of interfaces such as menus or question-asking. They are also a convenient framework for system extension facilities for use by the relatively casual user [10]; and they appear to be well-suited for building expert or assistant ones. By contrast, while they can be used to provide conversational or menu-driven environments, or those that require screen inputs, they tend not to work well, or at least to be difficult to support, when the underlying environment has any of those characteristics. These drawbacks impose some major constraints on system organization.

The difficulties and choices found in user communication with the system are paralleled in system communication with the user and with output presentation. The design of all formats around a 24 × 80 character display can be a severe limitation, especially for plots and graphics. Worse yet, in this day of stress on interaction, is the design of output for the line printer, with its headers and footers and its long and wide pages at low resolution. At the same time, there are still many analysts who prefer working with piles of paper to sitting at terminals; articles and food tables will probably be published and used on paper for a long time to come. Designs that depend on high-resolution graphics devices will exclude many users who will be unable to pay the price of entry to the system. Almost any solution will either make some class of potential users very unhappy or will limit the groups and types of users who can be served. This makes it very important to make decisions early and with an understanding of whose happiness is being considered and whose is being sacrificed.

DATA REPRESENTATIONS

Most integrated environments make use of some kind of file system to retain information – a worksheet, a special system file, or even a full data-base management system. In addition to providing a compact way to save information during and between sessions, these files can be used as the mechanism by which all commands for users, other than those intended to read raw data into files and display the contents of files, communicate with each other and with the users. In other words, computational commands do not read, clean, or process raw files, nor do they print results. Having commands work this way ensures (given adequate data representations) that any command can use the outputs of any other appropriate commands as inputs. That level of compatibility will apply to commands written in the future, as the system is extended, as well as to those designed initially. This type of strategy is also complementary to agent strategies and to primitive tool-based systems (discussed below).[1]

As a system-building approach, such strategy has a long tradition in statistical and social science computation [8, 9]. At the same time, many users find it inconvenient (unless it is hidden) for trivial sets of operations. It also leads to inconvenience and unpredictability when one discovers, late in the life of a system, that the data representation forms are inadequate, that there is no mechanism for cleanly extending them, and that the only practical solution is to have some commands that simply print results. For example, some statistical systems in the recent past have run into major difficulties as the requirements of new or proposed procedures forced a choice between moving from columns and data matrices to symmetric matrices and multi-dimensional arrays on the one hand, and, on the other, deciding that some routines should display results that could not be captured in the file system. We are aware of several situations in which systems have been reorganized in major ways internally, requiring users to convert data sets, in order to try to cope with these problems as they unfold. Naturally enough, the problems tend to be buried as much as possible, rather than being cited explicitly in the literature.

To a degree, the more heavily the system relies on a single fixed set of data structures, the more dependent it becomes on the correctness of those data structures and file representations; such dependence amounts to a negative technical aspect of the approach. So once again there is a challenge in trying to make the right decision – in balancing the compatibility advantages against convenience for trivial tasks and against the risks of having to use a mixed strategy or make a major redesign if the data representations are not adequate to future developments. There are some alternative methods for data conversion, such as globally changing all files that would not exist in an integrated environment. However, the risks of making a conversion of broad scope would be a threat to the integrity of multiple dependent programs operating off a common data base or data representation, as well as to a more structured system.

SYSTEM ARCHITECTURE AND LINKAGES

As systems become large, basic issues of organization assume an importance of their own. As long as the total size of the system is much smaller than the effective size of the machine, one can be fairly cavalier about simply putting things together and running them. But with very large systems new issues arise as to how the codes are to be accommodated on the machine. Depending on the size and characteristics of both the system and the machine, the consequences of trying to ignore the impact of size may be as gross as vastly increased costs or inability to run at all, or as subtle as slightly reduced performance or response. Independently of what the host machine or operating system looks like, it is generally desirable to keep as low as possible

---

1. The "pipe" approach used in the UNIX system is an alternative, but a less desirable one. Using the notion of files described here permits "strongly typed" and complex representations, whereas pipes do not.

the host system's perception of how large the applications system is in operation (the working set). At the same time, the user's perception of application system size should be one of a vast and growing collection of capabilities and functions. While the user will be happy to see more facilities in the system, that happiness will disappear quickly if the presence of additional facilities causes last year's task to operate more slowly or at higher cost. A system may be thought of as having two separate "sizes," one determined by the machine resources required to execute its individual programs, and the other determined by its total extent when stored. If a system has many and diverse capabilities, no single user is going to use more than a small fraction of the system on any given day or even in one lifetime. As a result, the first of these sizes must depend on what the user is doing in a particular session – the programs being run – rather than on the total system size (the second type of measurement). Very few operating systems provide facilities that make this easy, and simulating it in an application is quite difficult. An application can simulate it, however, and a few existing systems and packages have managed to do so, but it is not to be taken lightly and requires an intimate understanding of the operating environment that must be tricked.

An application that provides linkage facilities that depend only on what programs the user asks for, as they are asked for, has several advantages in addition to keeping the bills down, the performance up, and the machine requirements at a minimum for the user's application. Among the most important of these is the ability to utilize user-supplied codes to supplement system facilities without having to create private copies of the system or major components of it. Such copies are a liability, not merely because of the costs or inefficiency they entail and the burdens they place on the user, but because they promote retention of ancient versions of systems and codes and subsequent conversion problems.

Very large virtual memories may provide some of the facilities that will help in tricking an operating system into behaving well when confronted with large systems, but they are not in themselves a solution. Misuse of such memories can introduce severe performance degradation. The major benefit of large virtual memories, in addition to programming convenience, is that as application size limits are reached the system degrades more or less gracefully; in the absence of virtual memories, the same situation leads to catastrophic system collapse. The goal and the challenge should be to avoid both alternatives, especially when the amount of software and data actually being used fall well within the host system's natural limits for application sizes.

STABILITY

The issue of stability is very important if the system is to be large and to have a long life expectancy. The problem is similar to that of advanced operating systems: the developers and their staff want things to be better, current, and as sophisticated and clever as possible; programmers want a completely stable interface (this is especially true of those amateur users who write a little code to add one small facility to the system to make it perfect for their purposes). Users also believe that they want commands and results that are predictable: what worked last year should work today in exactly the same way. On the other hand, they also want the programs, algorithms, functional capabilities, and data to reflect last month's journal article. Of course, no one but developers sees incompatibilities among these objectives.

220

An obvious possibility is to make a rule that a routine, once installed under a particular name, has the same behaviour forever, and that new things or versions of things must have new names. This is an easier approach to apply to a subroutine library – especially a subroutine library from which users are permitted to copy and imbed "obsolete" routines that have been replaced by newer ones with different names[1] – than to a major integrated collection, where such an approach requires a great deal of discipline on the part of the maintainers and tolerance on the part of the users. Unless the architectural and linkage problems have been solved exceptionally well, the accumulation of semi-obsolete codes will also cause degradation and high disk costs much sooner than if the growth of the system resulted from an extension and replacement strategy alone.

While the stress here has been on programs rather than on data, the problems with data are quite similar. The ability to reproduce an old result may require that the associated data be retained forever (or nearly so), even when it has been found to be out of data or substandard. There have been a few cases in which users have rather carefully built procedures to compensate for data inadequacies that they were aware of, only to have their procedures produce seriously incorrect answers when the original data values were replaced with "corrected" ones. There are strong arguments for being able to associate particular vintages of data with particular analyses or studies, but such requirements impose great difficulties on the system's design and its human managers. So the data and program problems are not very much different after all.

PRIMITIVE TOOL-BASED SYSTEMS

Many of the problems – with programs if not with data – that have been discussed here can be avoided by designing a system around primitive tools that provide no more facility than what is necessary for a user to put the things together to produce the computations needed. Such a system provides adequate facilities for the right user, tends to be very extensible, and can typically be kept very small in spite of being integrated and powerful. Most important, such systems are conceptually very simple. However, they do tend to be disastrous for unsophisticated users and even sophisticated users spend too much time fussing around with the tools themselves. In a rich environment, that fussing often has more to do with the process of moving objects back and forth – looking for tools to make square pegs fit round holes – than with anything substantively interesting. Further, all other things being equal, systems of primitives tend to be slower in operation than higher-level integrated systems, and are sometimes so slow as to result in poor response rather than merely poor resource consumption. None the less, such an architecture may be a reasonable choice for some audiences.

---

1. Giving improved routines different names from the incompatible ones they replace has, for example, been a long-term IMSL strategy [13].

SUMMARY

The critical challenge in developing a large integrated system is the same as the critical challenge in developing any system: figuring out what the real goals and requirements are, and what is to be sacrificed in order to get there. The costs, complexities, and additional problems that arise when systems become very large, or when the integration requirements become more stringent, are sufficiently major that the decision to build such a system should be justified on the basis of real requirements.

Once the decision to build a large and integrated system has been made, and the necessary resources for planning, building prototypes and tools, examining design issues, and actually building the system have been secured, the short-term implications of a variety of questions that really make sense only in the long term must be considered. Each of these questions and issues poses a signficant challenge for which there are no clear answers that are right for all cases. Those discussed in this paper that directly affect the development process itself include the choice of the operating base, the language of implementation, organization of the system, and how stability and growth are to be managed and accommodated over time. More user-related issues include documentation, command languages, presentation and output, and how to permit users to extend the system when needed. A final design concern is how to contain damage resulting from incorrect decisions, which are inevitable no matter how much care is taken. The challenges are nearly overwhelming, but can and must be met.

REFERENCES

1. American National Standards Institute, *Proposed American National Standard Programming Language BASIC*, ANSI BSR X3.113-198X, 1983.
2. American National Standards Institute, *American National Standard Programming Language COBOL*, ANSI X3.23-1974. Revision (ANSI BSR X3.23-198X, June 1983; also ISO DP 1989.2) undergoing public review, autumn 1983.
3. American National Standards Institute, *American National Standard Programming Language PL/I*, ANSI X3.53-1976. Equivalent document approved by ISO, as ISO 6160-1979.
4. American National Standards Institute, *American National Standard for Minimal BASIC*, ANSI X3.60-1978.
5. American National Standards Institute, *American National Standard Programming Language PL/I General Purpose Subset*, ANSI X3.74-1981. Equivalent document under ISO review as ISO DP6522.
6. American National Standards Institute, *American National Standard Programming Language FORTRAN*, ANSI X3.9-1978. Equivalent document approved by ISO, as ISO 1539-1980.
7. K. N. Berk and I. Francis, "A Review of Manuals for BMDP and SPSS," *J. Amer. Stat. Assoc.*, 73, 361: 65–71 (1978).
8. S. Buhler and R. Buhler, *P-STAT 78 User's Manual* (P-STAT Inc., Princeton, N.J., 1979).
9. G. J. Culler and B. D. Fried, "The TRW Two-Station, On-Line Scientific Computer: General Description," in M. A. Sass and W. D. Wilkinson, eds., *Computer Augmentation of Human Reasoning* (Spartan Books, Washington, D.C., 1965), pp. 66–67.
10. R. Dawson and J. C. Klensin, "User Extensions to Statistical Software," in American Statistical Association, *1980 Proceedings of the Statistical Computing Section* (American Statistical Association, Washington, D.C., 1980), pp. 332–334.
11. US Department of Defense, *Reference Manual for the Ada(R) Programming Language*, ANSI/MIL-STD-1815A-1983. (Ada is a registered trademark of the US Department of Defense).
12. I. Francis, R. Heiberger, and P. Velleman, "Criteria and Considerations in the Evaluation of Statistical Packages," *Amer. Stat.*, 29: 52–56 (1975).

13. International Mathematical and Statistical Laboratories, Inc., *IMSL Library 2 Reference Manual* (IMSL, Houston, Tex., 1975).
14. International Standards Organization, *International Standard Programming Language – Pascal*, ISO 7185, 1983. Also published in the US as ANSI/IEEE770X3.97-1983 and in the UK as BS 6192-1982.
15. J. C. Klensin, "Short-term Friendly and Long-term Hostile?" paper presented at the Conference on Easier and More Productive Use of Computer Systems, 20–22 May 1981, Ann Arbor, Mich., and reprinted in *SIGSOC Bulletin*, 13, 2–3: 105–110 (1982).
16. M. R. Muller, "A Review of Manuals for BMDP and SPSS," *J. Amer. Stat. Assoc.*, 73, 361: 71–80 (1978).
17. Numerical Algorithms Group, *NAG Fortran Library Manual, Mark 10* (Numerical Algorithms Group, Oxford, 1983).

Participants

Lenore Arab
Klinisches Institut für
  Herzinfarktforschung
Heidelberg, Federal Republic of
  Germany

George H. Beaton
Department of Nutritional Sciences
Faculty of Medicine
University of Toronto
Toronto, Canada

Gary R. Beecher
Nutrient Composition Laboratory
Beltsville Human Nutrition Research
  Center
Beltsville, Maryland, USA

Lena Bergstrom
National Food Administration
Uppsala, Sweden

Åke Bruce
National Food Administration
Uppsala, Sweden

Ritva R. Butrum
Diet and Cancer Branch
Division of Cancer Prevention and
  Control

National Cancer Institute
Washington, D.C., USA

I. Marilyn Buzzard
Nutrition Co-ordinating Center
Division of Biometry
School of Public Health
University of Minnesota
Minneapolis, Minnesota, USA

Maria A. S. C. Coelho
Laboratory of Nutritional Biochemistry
Department of Nutrition
Federal University of Pernambuco
Cidade Universitaria, Brazil

Diane Feskanich
Nutrition Co-ordinating Center
Division of Biometry
School of Public Health
University of Minnesota
Minneapolis, Minnesota, USA

Hernando Flores
Laboratory of Nutritional Biochemistry
Department of Nutrition
Federal University of Pernambuco
Cidade Universitaria, Brazil

Margaret F. Gloninger
Department of Epidemiology
Graduate School of Public Health
University of Pittsburgh
Pittsburgh, Pennsylvania, USA

R. Gaurth Hansen
Department of Nutrition and Food
  Sciences
Utah State University
Logan, Utah, USA

Frank N. Hepburn
Nutrition Monitoring Division
Human Nutrition Information Service
US Department of Agriculture
Washington, D.C., USA

Joanne M. Holden
Nutrient Composition Laboratory
Beltsville Human Nutrition Research
  Center
Beltsville, Maryland, USA

Loretta W. Hoover
Department of Human Nutrition, Foods,
  Food Systems Management
University of Missouri–Columbia
Columbia, Missouri, USA

John C. Klensin
Principal Research Scientist
INFOODS Secretariat
Massachusetts Institute of Technology,
Cambridge, Massachusetts, USA

Hyun Kyung Moon Lee
Department of Epidemiology
Graduate School of Public Health
University of Pittsburgh
Pittsburgh, Pennsylvania, USA

Betty B. Peterkin
Human Nutrition Information Service
US Department of Agriculture
Washington, D.C., USA

Grace J. Petot
Departments of Nutrition,
  Epidemiology, and Biostatistics
School of Medicine
Case Western Reserve University
Cleveland, Ohio, USA

William M. Rand
Executive Secretary, INFOODS
INFOODS Secretariat
Massachusetts Institute of Technology
Cambridge, Massachusetts, USA

Anita Schubert
Nutrient Composition Laboratory
Beltsville Human Nutrition Research
  Center
Beltsville, Maryland, USA

Noreen B. Schvaneveldt
Department of Nutrition and Food
  Sciences
Utah State University
Logan, Utah, USA

Ann Sorenson
Office of Assistant Secretary of Health
Office of Disease Prevention and Health
  Promotion
Department of Health and Human
  Services
Washington, D.C., USA

Fred H. Steinke
Nutritional Biochemistry Department
Ralston Purina Company
St. Louis, Missouri, USA

Aree Valyasevi
Institute of Nutrition
Mahidol University
Bangkok, Thailand

Pamela C. Verdier
Bureau of Nutritional Sciences
Food Directorate, Health Protection
  Branch
Health and Welfare Canada
Ottawa, Ontario, Canada

Carol T. Windham
Department of Nutrition and Food
  Sciences
Utah State University
Logan, Utah, USA

Wayne R. Wolf
Nutrient Composition Laboratory
Beltsville Human Nutrition Research
  Center
Beltsville, Maryland, USA

Catherine E. Woteki
National Center for Health Statistics
US Dept of Health and Human Services

Hyattsville, Maryland, USA

Bonita W. Wyse
Department of Nutrition and Food
  Sciences
Utah State University
Logan, Utah, USA

Vernon R. Young
Laboratory of Human Nutrition
Department of Applied Biological
  Sciences
Massachusetts Institute of Technology
Cambridge, Massachusetts, USA